T0318381

Eco-Cities and Green Transport

Eco-Cities and Green Transport

Huapu Lu
Tsinghua University, Beijing, P.R. China

ELSEVIER

Elsevier
Radarweg 29, PO Box 211, 1000 AE Amsterdam, Netherlands
The Boulevard, Langford Lane, Kidlington, Oxford OX5 1GB, United Kingdom
50 Hampshire Street, 5th Floor, Cambridge, MA 02139, United States

British Library Cataloguing-in-Publication Data
A catalogue record for this book is available from the British Library

Library of Congress Cataloging-in-Publication Data
A catalog record for this book is available from the Library of Congress

ISBN: 978-0-12-821516-6

For Information on all Elsevier publications
visit our website at https://www.elsevier.com/books-and-journals

Publisher: Joe Hayton
Acquisitions Editor: Brian Romer
Editorial Project Manager: Naomi Robertson
Production Project Manager: Punithavathy Govindaradjane
Cover Designer: Matthew Limbert

Typeset by MPS Limited, Chennai, India

Working together
to grow libraries in
developing countries

www.elsevier.com • www.bookaid.org

Contents

Preface

With the rapid development of urbanization and motorization, China has entered a new era of forging ahead with eco-cities, new urbanization, and green transportation system construction, during which opportunities and challenges coexist. For a modern society with highly developed science and technology and material progress, the questions about what kinds of cities to build, what kinds of transportation systems to build, and what kinds of living environments to create are directly related to use of the earth's energy and resources, maintaining a good ecological environment, and achieving sustainable development. This is a golden age when cities and transportation infrastructures are constructed on a large scale, with the opportunity to build efficient and convenient, livable, safe, and healthy cities, that also have excellent ecological qualities and provide welcoming people-friendly environments. How can this goal of green transportation and eco-city development be achieved? These questions above have been researched for a long time, but they urgently need to be answered and solved in the field of urban transportation planning and construction in China.

Advice from others may help in answering these questions. The author and his team have studied cases of foreign cities in depth and systematically for over a decade with academic exchanges and special research. By learning from analytical summaries and comparative studies of other countries, they hope to provide experience and references for eco-cities and green transportation systems constructed in China.

Since the first city was created, people have been in constant pursuit of a more convenient, more comfortable, safer, more livable, and better urban environment. Ebenezer Howard, a British social activist, published an excellent book with far-reaching influence, *To-morrow: A Peaceful Path to Real Reform*, in 1898, the title of which changed to *Grande City of Tomorrow* in the second edition published in 1902. Howard had experienced significant changes to the urban landscape and urban life during the course of capitalist industrialization in the Industrial Revolution. What's more, he saw the many problems of urban expansion, such as excessive population concentration, environmental pollution, a wide poverty gap, and urban slums, his answer to which was the garden city. He envisaged future cities that not only experienced prosperity and convenience, but also blue skies with white clouds, fresh air, quiet woods, and green glass, as found in rural areas. The city is a place where people can work efficiently, but what is more important is that it should be a livable environment with an elegant landscape and ecological harmony that offers its inhabitants a healthy and relaxing life.

In 1933 the Athens Charter proposed that cities should be studied as a whole, including their surrounding areas. It was pointed out that the key for urban planning was to balance the four functions of the city: housing, work, recreation, and transportation. In addition, the importance of a better urban living environment was emphasized and the concept of urban planning considering functional zoning was proposed.

In 1978 the Charter of Machu Picchu comprehensively summarized the lessons learned from urban development since the publication of the Athens Charter. On the basis of confirming most of the principles of the Athens Charter, it highlighted that it should be endeavored to create a comprehensive multifunctional living environment rather than the excessive pursuit of functional zoning that resulted in the sacrifice of the organic organization of a city. The idea of transit-oriented transportation and paying attention to the development of transportation while considering environmental and energy issues was then proposed.

In 1999 the 20th World Congress of Architects adopted the Beijing Charter, drafted by Professor Liangyong Wu, a member of the China Academy of Sciences and China Academy of Engineering, and also a winner of the Highest Science and Technology Award. Based on the generalized architecture and sciences of human settlements theory, the Beijing Charter advocated all-round development integrating architecture, landscape, and urban planning. The Beijing Charter is a program of action guiding urban and rural construction in the 21st century, thus establishing our architects and planners' leading positions in the field of architecture and urban planning in the world today.

Due to the large scale of the urban system and the complex influencing factors, it is difficult to carry out large-scale experimental research in a city. However, the continuous exploration and practice of outstanding cities around the world has provided us with practical experience in urban transportation planning and construction. To summarize, by referring to these cases, clarifying their development background and experience characteristics, utilizing their full ideological essence, development goals, ways of implementation, and policy orientation, we are better able to realize our ambition of planning a new type of urbanization with eco-cities and green transportation systems. Therefore this book highlights the analysis of foreign cities in a development context, including their natural environments, traffic demand characteristics, planning and construction processes, and policy backgrounds. These can enlighten us, enabling us to think deeply and thoroughly understand the importance of the disciplines and goals of planning, the impact of traffic planning to city development, and the correlation among city, transportation, energy and environment, ecology, and living quality. On the one hand, this can help us to draw fully on international experience and on the other, it helps us avoid blindly copying or incomplete imitation. What should be clarified is that not all of these case studies are perfect and not all are eco-cities following the concept of green transportation. There are advantages as well as disadvantages in these, and the experience to be gained includes lessons of what to avoid in some cases. However, as these cities have accumulated experiences and developed with special features, they can inspire us to build upon their good work.

In short, the case studies provided in this book are not merely sources for imitation, their relevance more significantly lies in stimulating our innovative thinking, broadening our horizons, and helping us to build more efficient, convenient, ecological, healthy, livable, and beautiful cities in China. It is hoped that this book can be a reference for my friends and colleagues in the field of city planning, and I invite readers to submit their comments and criticisms.

As this book is about to go to press, I would like to express my sincere gratitude to Ms. Wenjie Zhao of Barton Wilmore, the United Kingdom, who arranged intensive visits to six European countries to facilitate our investigations, and Ramboll Company, the United Kingdom, and Letchworth Heritage Foundation, which provided some of the case studies presented herein.

Thanks are also expressed to Professor Yoshitsugu Hayashi of Nagoya University, Professor Bush of Munich Technical University, Professor Robot Cevero of University of California, Berkeley, Professor Ben Akiva of Massachusetts Institute of Technology, and Loh Chow Kway, Dean of Singapore Urban Transportation International Department, and many colleagues, who are too many to mention, for your help in the investigation and research into these case studies over the years.

In addition, some of the city density data and transportation mode rates were provided by Professor Yoshitsugu Hayashi of Nagoya University, Professor Haniwa Kim of South Korea, Dr. Ilina Irina of the University of Amsterdam, Dr. Rau of the Technical University of Munich, Dr. Doulet in France, Dr. Nelson of the University of Paulo Sao, Brazil, and Dr. Jiangping Zhou of the State University of Iowa, the United States, and I express my deep appreciation to all of you. Thanks also to Professor Liren Duan for the invaluable photos of the changes to Seoul City Hall Square and the traffic space configuration of the Han River Diving Bridge.

Finally I would also like to thank Dr. Jing Yang (cases about Rio de Janeiro, Carmel, San Carlos), Dr. Yu Ding (Paris, Malmö, New York, Los Angeles), Dr. Zhiyuan Sun (London, Letchworth), Dr. He Ma (Copenhagen, Madrid), Mr. Wenbo Kuang (Stockholm, Amsterdam, Seoul, Tokyo), Dr. Jin Wang (San Francisco), Dr. Yang Lu (Singapore), Dr. Jing Wang (Curitiba), and Pei Su (St. Petersburg, Munich), for your help with the first draft of this book.

Huapu Lu
Tsinghua University, Beijing, P.R. China
August 1, 2019

Chapter 1

Copenhagen, Denmark

Chapter Outline

1.1 Overview of the city

Copenhagen is the capital, largest city, and largest sea port of the Kingdom of Denmark, and is also the largest city in Nordic countries. It is the home of the famous fairy-tale writer Andersen, a historical and cultural city, and the political, economic, and cultural center of Denmark. Hovedstaden is located in eastern Denmark, including Copenhagen city, Copenhagen county, Frederiksberg city, Frederiksberg county, Roskilde county, and Barnhdm region, etc. "Copenhagen city center" refers to the central city of Copenhagen (i.e., the old town), that is, the "Indre By" part of Fig. 1.1. "Copenhagen city" refers to the area of Copenhagen city center and its surrounding suburbs, all of which are shown in Fig. 1.1. "Copenhagen county" and "Copenhagen city" both belong to the region of Copenhagen, but do not

Eco-Cities and Green Transport. DOI: https://doi.org/10.1016/B978-0-12-821516-6.00001-1

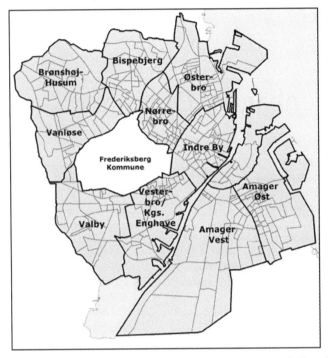

FIGURE 1.1   Map of Copenhagen city area. *From http://zh.wikipedia.org/wiki/Copenhagen.*

overlap in administrative division. Copenhagen county is an aggregation of other small towns. Copenhagen city is the most active and representative city in the region of Copenhagen. Without special explanation, "Copenhagen" refers to "Copenhagen city" in the common sense in this book [1].

The areas, populations, and densities of Copenhagen's regions are shown in Table 1.1. GDP statistics for different years in the region of Copenhagen are shown in Table 1.2.

1.2 Urban mobility development and motorization policy

Statistical data in 2010 show that the number of motor vehicles in Copenhagen was 2.1 million, with a private car ownership rate of 22.3%. According to the latest data from the Copenhagen Police Bureau in 2012, there were about 312,600 motor vehicles entering and leaving Copenhagen daily, accounting for only 14.9% of the total vehicle ownership in the city, which shows that the usage frequency is very low [2].

From 2000 to 2010, the number of motor vehicles in Copenhagen increased by 17% and the traffic volume in the region of Copenhagen increased by 8%, while the road traffic volume in the five basins area in the

TABLE 1.1 List of the populations, areas, and densities of Copenhagen [2].

	Region of Copenhagen	Copenhagen city	Copenhagen city center
Area (km^2)	2553.1	455.61	77.2
Population	1,736,889	1,230,728	562,379
Density (people/km^2)	680	2700	7280

Source: StatBank, Denmark. Data from http://www.statistikbanken.dk/BEF1A07.

city center decreased by nearly 10%. This is inseparable from Copenhagen's

TABLE 1.2 Changes to GDP in the region of Copenhagen [2].

Year	2005	2006	2007	2008	2009	2010	2011
Per capita GDP (ten thousand dollars)	6.32	6.47	6.69	6.86	6.58	7.14	7.11

advocacy of "Green Transportation—by Bicycles" and its restrictive policy on motor vehicles [3].

In terms of motor vehicle policy, Copenhagen mainly adopts a restrictive approach. Despite the small motor vehicle traffic volume, the government continues to impel people to abandon private cars and switch to green modes of transportation such as public transport, bicycles, and walking. The restrictions on motor vehicles in Copenhagen are mainly reflected in two aspects: car purchase and parking.

As early as World War II, the Copenhagen government imposed a ban on car imports and maintained it for quite a long time after the war. Since the late 1970s, the Copenhagen government has lifted the tax on private car purchase, with car taxes roughly three times the car price [1].

For parking, the city government has adopted a rather innovative and politically wise parking policy, which focuses on reducing parking spaces in the central urban areas. In order to avoid the strong opposition of car owners by reducing too many parking spaces at one time, the city government had been taking a gradual strategy in the past decades, insisting on reducing parking spaces by 2%–3% every year. This approach accumulated significantly reduced parking spaces and avoided aggravating social contradictions. In addition, the parking price is defined by zones, so that the price is linked to the supply and demand relationship of the parking area and the level of public transport service. Furthermore, the city government takes into account both the increase of parking price and the reduction of parking spaces, so as to maintain the parking vacancy rate in urban central areas at around 10%, avoiding complaints from car owners, and thus reducing the resistance to policy implementation [4].

1.3   Urban structure and land use

Copenhagen's urban land use and transport corridor structure is known as "Finger Planning." The "Finger Planning" principle was first put forward in 1947. The hand-shaped urban skeleton structure is formed with Copenhagen city as the center of the palm, and five finger-shaped axes extending to the north, west, and south. The wedge-shaped zones between the fingers serve as forest, farmland, and open leisure space. For more than half a century, Copenhagen has steadily improved city planning and construction following the "Finger Planning" principle. The formal "Five Planning" was formulated in 2007, clearly pointing out that the direction of urban development was to build rail transit from the center to the periphery along the finger directions, to improve public transport facilities, and to build residential areas along the transport corridor. The wedge-shaped zones between fingers do not incur construction and development of urban facilities, but maintain the ecological characteristics of forests and green spaces. The supplementary spaces needed for urban development and construction of city groups are taken from other areas [5]. The planning and construction of the city center (palm part) focuses on improving urban public transport and nonmotorized traffic service systems, such as walking and bicycles, in order to satisfy short-distance travel within the city center and the need for green and environmental protection. The urban periphery areas (finger parts) provide space for urban expansion and construction of the new town, while rail transit is the main transportation mode to the city center. The planning and construction of periphery areas focus on improving infrastructure construction and improving the level of public transportation services.

1.4   Characteristics of the mode split

Due to the effective implementation of "Finger Planning," most travel in Copenhagen relies on green transportation: long-distance travel mainly depends on rail transit and bus, while short-distance travel within the city mainly relies on walking and bicycles. According to the latest data from Statistics Denmark, the citizen travel mode split in Copenhagen in 2011 is as shown in Table 1.3.

Based on the data above, it can be seen that green transportation is the main mode of travel in Copenhagen, with private car travel accounting for

TABLE 1.3 Split of travel modes in Copenhagen (2011) [2].

Transportation mode	Walking	Bicycle	Train + light rail + bus	Private cars	Taxi
Mode split (%)	11	35	22	30	2
Total green transport (%)		68		–	–

30%. A higher modal rate of bicycle use and walking has promoted the healthy lifestyles of Copenhagen residents, and greatly promoted the formation of a green, ecological, and sustainable urban environment.

1.5   Urban roads and public transit

Rail transit is the dominant mode of transportation in Copenhagen. Due to the "Finger Planning" land use principle adopted by Copenhagen, the traffic corridors supported by the rail transit radiate from the city center to the peripheral areas in five directions. In the process of the city's extension to these peripheral areas, the development and construction of most infrastructure is concentrated near the railway stations, such as residential buildings and main roadways. Thus the formation of the "Finger Planning" is formulated. In line with this, the rail transit network also shows a hand (palm and fingers) shape that matches the urban layout, as shown in Fig. 1.2. It shows that the implementation of "Finger Planning" in urban planning and construction also promoted the integrated development of the transportation system and land use. At present, Copenhagen has nearly 200 km of railway, including 85 suburban railway stations and 35 urban railway stations. Rail transit undertakes about 500,000 commuter trips daily, accounting for about one-third of the total daily motorized trips. It undertakes the traffic demand of the corridor between the city center and the "finger" areas within a radius of 40 km [6].

Effective mixed development of land has been implemented along the railway and around stations during the construction of the rail transit system, so that people can quickly enter the business, shopping, and residential areas without transferring to other modes of transportation after exiting the station. Some station buildings themselves are comprehensive hubs, providing people with concentrated areas for travel, work, shopping, and living. This is why people tend to choose rail transit for personal trips. Copenhagen Central Railway Station, located in the city center, is a very old building, as shown in Fig. 1.3. The track lines pass under the building without affecting the

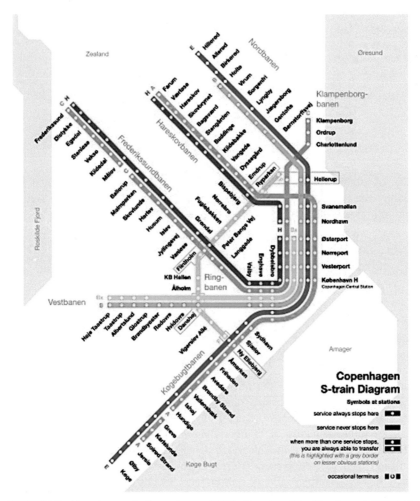

FIGURE 1.2   Copenhagen suburban train map (2011). *From http://zh.wikipedia.org/wiki/File:S-train_diagram_%28dec2011%29.svg.*

landscape. Inside the station, there are waiting halls leading to the platforms as well as a number of shops selling food and small commodities, offering convenient supplies for travelers including food and basic shopping needs.

Buses, an important part of the public transport, also take a large share of trips in Copenhagen. Customers use a "swipe card" to get on and off the bus. Bus stations provide rest benches, bus route signs, city bus network maps, and real-time arrival time of the next bus. Most buses are large, enabling more people to ride, including those carrying bicycles with them. This is convenient for people choosing green transportation.

FIGURE 1.3 Copenhagen Central Railway Station.

In terms of operation, the three major networks of the suburban railway system, urban rail transit, and urban bus are effectively combined. Unified tickets are provided although the three networks despite them belonging to different companies. The settings of bus stations and rail transit stations are also designed as a whole to facilitate the transfer of travelers.

1.6   Bicycle traffic system construction

Copenhagen has always been known as the "city of bicycles," which first benefits from the Danish people's overall advocacy for green, environmental, and fitness concepts. Denmark has a population of more than 53 million, and more than 3 million bicycles. Copenhagen, as the capital of Denmark, is a model area for bicycle advocacy. About one-third of commuters in Copenhagen travel by bicycle, including government officials, business tycoons, and celebrities. Second, the Copenhagen government has vigorously promoted bicycle traffic for a long time, increasing the bicycle lane infrastructure, and building a huge bicycle lane system with a total length of 350 km. In line with the urban structure of "Finger Planning," the regional bicycle lane system in Copenhagen has been formed in this similar shape. The Copenhagen government has constantly set new goals for promoting bicycle transportation. The goal for 2015 was that the modal share of bicycle would achieve 50% for working commuters and students [7].

To aggressively promote bicycle traffic, the Copenhagen government and relevant agencies have taken the following effective measures.

1.6.1 Construction of green corridors for bicycles

There are 40 km of bicycle green corridors in Copenhagen [8], which are completely separated from motor lanes, as shown in Fig. 1.4. When there is a bus stop in front of the road, the bicycle green passage is designed to turn naturally to the right, then around the back of the bus stop, finally arriving back to the left direction parallel to the motor vehicle lane after the bus stop, as shown in Fig. 1.5. This alignment design enables bicycles on the green

FIGURE 1.4 Target and right-turn signs on the bicycle green passage.

FIGURE 1.5 Alignment design of the green passage around bus stops.

passage not only to avoid affecting buses stopping and passengers getting on/off, but also to maintain the continuity, rapidity, and safety of bicycle traffic.

1.6.2 "Lifted" bicycle lanes

The construction level of Copenhagen's bicycle lanes is high and the maintenance system has been perfected and is well developed. As shown in Fig. 1.6, the pavement of bicycle lanes is lifted to 7−12 cm higher than that of vehicle travel lanes. The separation zone between the bicycle lane and vehicle lane is built with a curbstone, and there is an independent drainage system, which helps with the maintenance of bicycle lanes. The speed of bicycles on special roads can be comparable to that of cars. Pedestrians or other vehicles are not allowed to pass or stop on bicycle lanes, whose traffic priority is the same as that of vehicle lanes.

1.6.3 Dedicated signal lights and special waiting areas for bicycles

In Copenhagen, bicycles have the highest travel priority. Special waiting areas are set for bicycles at intersections. As shown in Fig. 1.7, pedals are also set up in some places to facilitate cyclists resting. In terms of signal configuration, longer green time is designed for bicycles: not only special signal lights are set for bicycles, but also bicycle "green wave" signal linkage control is adopted for roads with higher bicycle traffic. The roadway network makes bicycle traffic faster and more convenient.

FIGURE 1.6 Cyclists on bicycle lanes.

FIGURE 1.7 Cyclists waiting for signals at intersections.

FIGURE 1.8 Bicycle parking lot near a residential area.

1.6.4 Construction of bicycle parking spaces

In response to the growing demand for bicycle parking, the Copenhagen government has adopted a gradual increase method in the construction of parking facilities. For residential and office areas where the contradiction between supply and demand is prominent, owners are allowed and encouraged to reform their vehicle parking spaces into bicycle parking spaces, as shown in Fig. 1.8. For general streets, bicycle parking facilities are designed to be added to the corners of sidewalks. For commercial streets, where land is scarce and expensive, street-facing shopkeepers are allowed to use the remaining space of the sidewalk to build bicycle racks.

1.6.5   Integration of bicycles and buses

In order to support the mass use of bicycles, the connection between public transport and bicycles is almost perfect in Copenhagen. Near the bus stop, bicycle parking lots are well equipped to facilitate the transfer of cyclists taking buses. Bicycle placement devices are equipped on buses, so that bicycles can be taken by bus and rail transit, providing as much convenience as possible for cyclists.

1.6.6   Public bicycle rental system

Bycyklen (The City Bike), which has been in operation since 1995, is the main public rental bicycle project in Copenhagen. The project is operated jointly by the government and the private sector. It is open to the public and tourists free of charge. There are 110 bicycle rental stations and more than 2000 public bicycles in operation, and they have achieved good results [8].

For quite a long time, Copenhagen has aggressively promoted the advanced concept of bicycle transport, invested heavily in the construction of bicycle transportation infrastructure, and given preferential policies and high-level management to bicycle transportation. All these efforts have made bicycle riding in Copenhagen fashionable, creating a rich "bicycle culture." Whether it is sunny or rainy, cyclists' excellent skills, fast speed, and steadfast figure form a beautiful landscape, which inspires other people to also engage in the activity. Walking in the streets of Copenhagen, when you accidentally stray into a bicycle lane, the cyclist may roar past you just as you hear their crisp bell sound, soon disappearing at the end of the street. Copenhagen deserves the title of "Bicycle City."

1.7   Pedestrian traffic system in the old town

The old town of Copenhagen has a long history and rich cultural heritage. Many traditional residential and commercial buildings, and cultural facilities are centrally located there. Its population density is about 2.7 times the average population density of Copenhagen. In order to solve the traffic problems as well as maintain the traditional human landscape and ecological environment in old districts, green transportation modes such as walking, bicycles, and buses are actively promoted by the Copenhagen government. Pedestrian system construction is carried out in traditional old districts also.

In 1962, Copenhagen began to renovate its pedestrian-only streets and pedestrian-priority streets. Vehicle traffic was banned in the busy old streets, and the world's first commercial pedestrian-only street was built, namely, Strøget Street, as shown in Figs. 1.9 and 1.10, and Gammeltorv Square (Fig. 1.11). The commercial pedestrian-only street, which runs from the City Hall to the canal connecting the Baltic Sea, is lined with shops on both sides

FIGURE 1.9 Strøget Street, Copenhagen (1).

FIGURE 1.10 Strøget Street, Copenhagen (2).

of the street. These create a haven of leisure shopping for Copenhagen residents and tourists. For nearly half a century, Copenhagen has insisted on promoting its pedestrian transportation system, especially pedestrian system construction in the old town areas and commercial streets. In view of the characteristics of the narrow width and dense road network of the central street in the old town areas, the diversion of people and vehicles is implemented to form a pedestrian network in the central area of the city. Hence the historical design has been retained.

FIGURE 1.11   Gammeltorv Square, Copenhagen.

1.8   Exploration and innovation of the modern architectural design

In the process of moving toward becoming a modern city, Copenhagen, on the one hand, vigorously develops green transportation, and controls the development speed of motorization and the use of cars to within a reasonable range. On the other hand, Copenhagen pays attention to the construction of ecological cities, as well as the protection and inheritance of traditional buildings and cultural landscapes in old town areas. Innovative architectural designs are explored and new districts are developed with environmental protection, energy savings, and a beautiful environment, creating a livable city. The following cases are worth learning about.

1.8.1   8-TALLET

The 8-TALLET is located in the new area of Amieu Island, Copenhagen, adjacent to the canal system, and is a group of community residential buildings. Named after its "8" shape from the aerial view, the total area of residential buildings is 61,000 m^2, including 476 units whose areas range from 65 to 144 m^2. The design purpose is to be affordable for households with different levels of income, in order to promote mixed living of wage earners with different salaries. About 10,000 m^2 of shops and office areas are provided within the buildings. The 8-TALLET is a group of apartments for average wage earners. Designed by the Danish architectural firm BIG (Bjake Ingels Group), the building was named "Best Building" at the Barcelona World Architecture Festival in 2011 [9].

The 8-TALLET is an enclosure building group, with each household able to share the green landscape of the residential area (Fig. 1.12). An open-air corridor with a slight slope is designed from the first floor to the top floor. Residents do not need to take elevators or climb stairs. They can walk along the corridor or ride bicycles to all floors and directly to their homes, as shown in Fig. 1.13. There are small open-air gardens in front of each entrance, and the entrance is directly connected with the outer corridor.

FIGURE 1.12   Green landscape sharing in enclosure buildings.

FIGURE 1.13   Walking or cycling along the gentle ramp corridor to reach each floor.

FIGURE 1.14 Open-air gardens in front of each entrance, which is directly connected with the outer corridor.

This design provides every resident with the feeling of living on the ground floor and being in touch with nature at any time (Fig. 1.14). At the same time, it has increased the space and opportunity for residents to meet and communicate with each other, avoiding the drawbacks of modern people living in closed unit buildings with few personal exchanges. The delicate conception and humanized design highlight Danish people's respect for nature and full consideration of pedestrians and bicycle users. In addition, the 8-TALLET is a multifunctional mixed development community, which includes various functional spaces such as public rental housing, shops, and office space.

A rainwater collection system, a large reservoir, and river course have been built into the 8-TALLET. The rainwater collected forms natural lakes and rivers. This not only makes use of rainwater resources effectively, but also creates hydrophilic space and beautiful scenery for the community, as shown in Figs. 1.15 and 1.16.

1.8.2   Pursuing the individualized design of architecture

"No two identical buildings" is Denmark's unique architectural concept, and also a challenging requirement for architects. In addition to meeting people's functional needs and aesthetic pursuits, no building can simply repeat other architectural designs. In order to guarantee the novelty and uniqueness of architectural design, as well as the comprehensive performance of energy saving, environmental protection, and ecological resources, competition is usually adopted to selecting architectural design schemes. Because of this,

FIGURE 1.15 Waterfront housing with a rainwater lake.

FIGURE 1.16 Beautiful community environment with a rainwater river.

every building in Denmark has its own unique style, no matter whether it is on an old street or a new one, or if it is a modern residential area or a commercial pedestrian-only street. However, the overall landscape effect is very coordinated. As shown in Fig. 1.17, these two residential buildings in the same area have the same architectural style. The facades adopt different colors and the geometric lines of the balcony are slightly changed to avoid simple repetition and show a lively design. A single-color group of residential buildings is shown

FIGURE 1.17   Color changes are used to avoid simple repetition for neighboring buildings with the same architectural style.

FIGURE 1.18   Architectures of the same style and color are located high and low to avoid monotony.

in Fig. 1.18, which avoids monotony by means of high–low scattering and balcony position change. Fig. 1.19 shows a common residential building. The balcony is made of transparent glass, while it is designed as a triangular balcony instead of the conventional rectangular form. One would feel like standing on the deck of a ship when standing at the top corner. Meanwhile, it is a bold idea to stagger the upper and lower balconies so as to increase the permeability of the residents on the balconies.

FIGURE 1.19   Transparent glass, triangular balcony, similar to the deck of a ship.

FIGURE 1.20   The circular dormitory building of the University of Copenhagen.

The personalized design of each building is also pursued for public buildings in Copenhagen. The student dormitory building of the University of Copenhagen on Amieu Island is shown in Fig. 1.20. The circular architecture is inspired by the design of the Tulou dwelling house in Fujian Province, China. The wooden facade gives a simple and heavy landscape effect.

FIGURE 1.21   Copenhagen Concert Hall building.

The floor windows of each room can be fully illuminated to create a spacious, bright, and vibrant student accommodation space. The concert hall building is shown in Fig. 1.21, with no special features in shape for the large space and steel structure building. However, the translucent plastic material layer decorating the facade of the building can be replaced or opened. At night, different color effects can be achieved by lighting operations in the plastic layer. The famous Bella Sky Hotel is shown in Fig. 1.22. There are two 76.5 m high towers, bending in opposite directions with an inclination of 15 degrees, creating a lissome appearance. Each guest room is designed with an open view into the building. In order to increase the natural landscape of the building, plant walls (Fig. 1.23) are set up in the lobby entrance. Guests hence feel as though they were in the natural environment, and the building is closely integrated with the ecological nature.

1.8.3   Emphasis on landscape construction after deindustrialization

In order to build a city with a beautiful environment and natural ecology, great importance is attached to landscape construction and ecological restoration of industrial waste sites in Copenhagen. As shown in Fig. 1.24, a picturesque park was built on desolate industrial land after the industrial production function had been removed. Sports facilities, leisure seats, and entertainment space are provided for citizens. Every spring, cherry trees blossom on both sides (Fig. 1.25), creating an excellent place for citizens to spend their holidays and enjoy flowers.

FIGURE 1.22 Architectural shape of the Bella Sky Hotel.

FIGURE 1.23 Plant wall in the Bella Sky Hotel lobby.

1.8.4 Protection of the traditional architecture and humanistic landscape in the old town

The traditional architectural style of the old town has been well preserved while the new city is under construction in Copenhagen. The New Harbor Canal, built in the 17th century, connects the shopping street of the old town with the Erle Strait. The fairy-tale buildings, with an over 300-year-old

FIGURE 1.24   Park built on a desolate industrial site.

FIGURE 1.25   Cherry trees in the renovated park.

history, are well preserved on both sides of the canal, among which Andersen's former residence is located. Nowadays, the people of Copenhagen use the canal for water transportation and sightseeing. There are many yachts and coffee tea-houses along the canal bank, giving tourists a picturesque feeling. Here, people can enjoy coffee and gourmet foods, while enjoying the fairy-tale scenery on both sides of the canal, as shown in Fig. 1.26.

FIGURE 1.26 Traditional buildings are on both sides of the New Harbor Canal.

1.9 Summary of the green transport and eco-city construction experience in Copenhagen

Copenhagen enjoys the reputation of "the most livable city" in the world. Summarizing the characteristics of Copenhagen in promoting green transportation and eco-city construction, the following experiences are worth learning from.

1. *Persisting in the guiding role of planning for urban development over the long term*

   The "Finger Planning" principle of Copenhagen was put forward in 1947. For more than half a century, the core idea of the original plan has not been changed, irrespective of the economic situation is or the political leaders. In Copenhagen, urban development is guided by transportation planning, which forms a good urban structure and land use pattern, and greatly promotes the formation and development of green transportation. Copenhagen is an excellent model for the integration of transportation and land use development and the construction of an ecological city.

2. *Taking comprehensive measures to promote the development of green transportation*

   First, the mixed development of urban land is adopted along rail transit lines, effectively promoting rail transit as a long-distance travel mode for residents and reducing the frequency of private car use. Second, bicycles are vigorously advocated, with heavy investment in infrastructure construction for the bicycle transportation system, and many preferential policies for cyclists in transportation management. In aspects of cultural fashion, the formation of a "bicycle culture" has promoted these efforts. Third, the construction of a pedestrian system in traditional districts is emphasized, creating a friendly environment for walking. Comprehensive measures have made Copenhagen a modern city dominated by green transportation.

3. *Pursuing architectural art, paying attention to environmental protection and energy conservation, and promoting the construction of an eco-city in every aspect*

Pursuing novel architectural designs, vigorously advocating symbiosis with nature, attaching great importance to environmental protection, and energy conservation have a huge part in the urban development of Copenhagen. While developing and building new cities, Copenhagen pays attention to protecting traditional buildings and old city blocks, presenting a people-oriented, sustainable, and ecological city example.

References

[1]   Copenhagen, Wikipedia, <http://en.wikipedia.org/wiki/Copenhagen>.
[2]   Statistics Bureau of Denmark, <http://www.statistikbanken.dk/BEF1A07>.
[3]   Jingzhong Y, Bo W. Copenhagen: green transportation and avoiding congestion. Econ Ref 2012;10.
[4]   Jun F, Kangming X. Research on TOD mode in Copenhagen. Urban Transp China 2006;2:41−6.
[5]   Xiaoping Y, Jianrun C. Inspiration of "Finger Planning" in Copenhagen. Urban Transp China 2011;9:71−4.
[6]   Meng Y. "City on Wheel" or "City on Rail"? Inter Des 2010;5:24 + 41−44 + 38.
[7]   Li W. Bicycle transport policy of Copenhagen. Beijing Plan Rev 2004;2:46−51.
[8]   Yang J, Yulin C, Yuanling Z, Jia X. Cycling revitalization strategies in cities under the context of motorization: a case study of Copenhagen. Mod Urban Res 2012;9:7−16.
[9]   8-TALLET official website, <http://www.8tallet.dk/>.

Further reading

Hongbin Z. Building ideas on public space of Copenhagen. Urban Probl 2011;9:81−9.

Chapter 2

Stockholm, Sweden

Chapter Outline

2.1 Overview of the city

Located on the west bank of the Baltic Sea, Stockholm is the capital of Sweden and its largest city. The Stockholm metropolitan area consists of

Eco-Cities and Green Transport. DOI: https://doi.org/10.1016/B978-0-12-821516-6.00002-3
25

26 cities, of which Stockholm is located in the central part. It is the most densely populated and economically active city and the political, economic, and cultural center of Sweden. Table 2.1 shows the population, area, and population density data for Stockholm and the Stockholm Region. According to the table, the population density in the urban area of Stockholm is 14 times the population density of the Stockholm region.

The city of Stockholm is spread over 14 islands and a peninsula. Parks and green areas account for 33% of the city's area. The city, which is known as the "Venice of the North," is surrounded by water with a healthy ecological environment. Fig. 2.1 shows the distribution of green belts and built-up

TABLE 2.1 Population, area, and population density in the Stockholm metropolitan region and Stockholm city (2012) [1].

	Stockholm region	Stockholm city
Population	2,127,006	881,235
Area (km^2)	6526	188
Population density (people/km^2)	326	4687

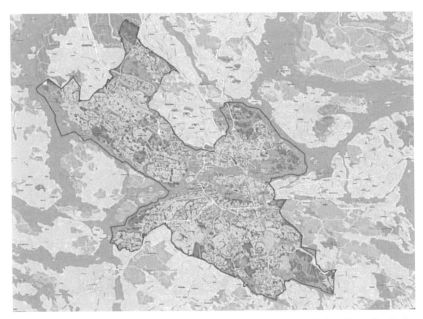

FIGURE 2.1 Distribution of green belts and built-up areas in Stockholm. *From Stockholm: European Green Capital 2010.*

FIGURE 2.2 Beautiful buildings on Riddarholmen Island in Stockholm. *From Stockholm: European Green Capital 2010.*

areas in Stockholm. There are many islands throughout the blue sea, with beautiful buildings on the islands. The green trees and surrounding blue waters form a beautiful picture (Fig. 2.2).

2.2 Urban structure and land use

Stockholm has distinctive urban planning and construction features. After World War II, Stockholm focused on a public transport-oriented land use model and a suburban development strategy of "large dispersion and small concentration" in the urban master plan [2], building several new towns around the central city. These new towns, which have independent city functions, are separated by open ecological corridors. Therefore the urban development model can be seen as "central city + new towns," as shown in Fig. 2.3. Under the division of the ecological corridor, all parts of the city are surrounded by green belts. At present, half of the population lives in the central city, with the other half in the new towns. The new towns are connected to the city center through a fast, radial regional-level rail system [4]. On the other hand, the traffic development model provides effective support for urban land development along the rail transit route.

The construction of new towns in Stockholm can be roughly divided into two phases. The first generation of new towns was typical ABC towns (A = housing, B = jobs, C = services), combining residence, employment, and service function. At the same time, the "half and half" population and employment planning principles were proposed, that is, half of employed people live in the new town, with half of its residents also working in the

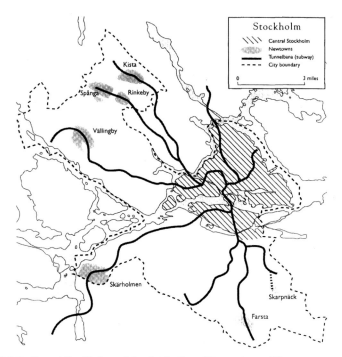

FIGURE 2.3 Central Stockholm and the distribution of its new towns [3].

new town. In the development process of the first generation of new towns, there were two problems. First, although the ratio of employment to resident population in the first generation of new towns reached 1.02, only one-third of the residents worked in these new towns, which did not meet the expected goals. Second, the design of the new towns is similar. The second generation of new towns replaced the concept of realizing the balance of living and employment within an independent new town by achieving an overall balance between all new towns [4]. At the same time, it presents a variety of characteristics in urban planning and construction.

2.3 Urban traffic system and traffic demand characteristics

In 2012, there were 318,131 motor vehicles in Stockholm, equating to 361 vehicles per 1000 people [1]. In terms of quantity, the level of motorization in Stockholm is relatively high, but from the perspective of its development of motorization, Stockholm's public transport-led development is worth learning from. From 1980 to 1990, per capita motor vehicle travel distance decreased by 229 km/year. The number of motor vehicles held by 1000 people remained stable, but the proportion of public transport use continued to grow [5]. This is unique in the context of the rapid increase in motor travel around the world.

There is a large demand for transportation between the central city of Stockholm and the new towns. In October 2005, the average traffic volume in and out of the downtown area reached 528,000 vehicles per working day [6]. Since August 2007, in order to alleviate the traffic congestion in the central city and reduce the environmental pollution caused by transportation, Stockholm has implemented a congestion fee system in the central city. The congestion fee area is controlled by 18 tollgates. Fig. 2.4 shows the scope of the congestion fee in Stockholm. The fee collection period is from 0630 to 1830 on weekdays. According to different time frames when entering the central city, each travel into the congestion fee area costs 10−20 Swedish krona.

The implementation of the congestion fee system in the central city has changed residents' mode of choice of travel. The public transport sharing

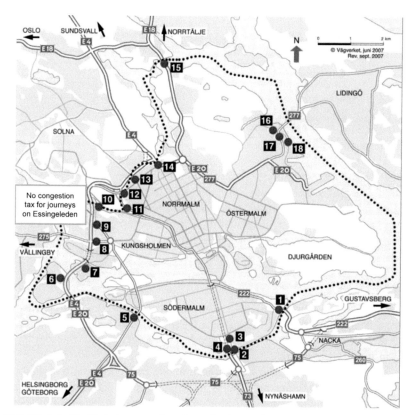

FIGURE 2.4 Stockholm's congestion fee collection region. *From Analysis of traffic in Stockholm with special focus on the effects of the congestion tax.*

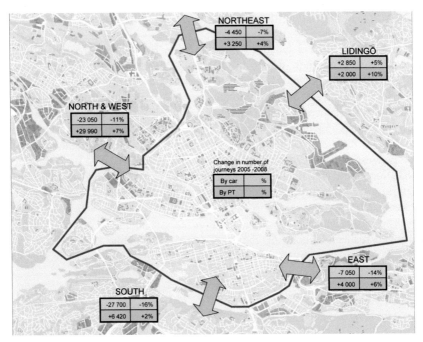

FIGURE 2.5 Changes in traffic sharing rates before and after congestion charges (fall 2008 vs fall 2005). *From Analysis of traffic in Stockholm with special focus on the effects of the congestion tax.*

rate in travel to the central city during the working day has generally increased by 7% [7]. Fig. 2.5 shows the specific data on the changes in the travel-sharing rate for cars and public transport entering and leaving the central city in different directions. The effect of the congestion fee collection system has gradually changed people's perceptions of the congestion fee. After several years of implementation, the overwhelming majority of the public have accepted and welcomed this policy.

2.4 Public transportation

Stockholm is the national railway center in Sweden, with the central railway station being the busiest railway station in the Nordic region. Stockholm is also the hub of the Nordic highway, and has the largest public transport system in the Nordic region.

Stockholm has developed an advanced public transport system. Table 2.2 lists the status of the Stockholm public transport system. There are 3 subway lines and 105 subway stations in Stockholm, with a total length of 108 km. There are 4 commuter rail lines and 78 stations with a total length of 257 km. There are 9 light rail lines and 110 stations with a total length of

TABLE 2.2 Stockholm public transport system status [8].

Type	Condition
Subway	Three groups of lines, 105 stations with a mileage of 108 km; average station spacing 1.03 km; passenger turnover of 1.796 billion person-km, annual passenger traffic of 320 million passengers, line load intensity of 0.81 million person-km/day (2012)
Bus	Passenger turnover of 1.83 billion person-km (2012)
Commuter railway	Four lines, 78 stations, mileage 257 km, average station spacing 3.29 km; passenger turnover 13.37 billion person-km, annual passenger traffic 76 million passengers, line load intensity 0.08 million person-km/day (2012)
Light rail	Nine lines, 110 stations, mileage 113.1 km; average station spacing 1.03 km; passenger turnover 278 million person-km, annual passenger traffic 45 million passengers, line load intensity 0.11 million person-km/day (2012)

113.1 km. Transferring between the subway and other modes of transportation is convenient [9]. City metro, suburban railway, light rail, and bus services are operated by Storstockholms Lokaltrafik (SL).

The service level of public transport in Stockholm is very high, and mainly reflected in the four aspects: departure frequency, on-time rate, in-vehicle facilities level, and station coverage.

First, the departure frequency of the Stockholm subway is about 6 min, the bus frequency is about 8 min, the frequency in nonpeak hours is about 15 min, and the buses operate till 2 a.m., with convenient frequency and service time. Second, all buses depart and operate on time. The punctuality rate is high, with each station having a bus schedule stating the bus arrival time. Third, public transportation facilities are advanced, the load intensity is not high, and buses are comfortable. Fourth, the station coverage is high. The distance between metro stations in Stockholm city is only 1 km, and the bus stations are more dense. In the city, a bus station can be located within no more than 300 m. It is very convenient for people to travel by public transport.

Therefore, the share of public transport in Stockholm is high. Fig. 2.6 shows the share of commuter traffic in Stockholm according to whether they live and work in the same part of the city. It can be seen from the data in the figure that more than 60% of commuters who work and live in different parts of the city use public transportation to travel to and from work. The public transport share exceeds 61%. Commuters living and working in the same new town are more likely to walk or ride a bicycle to and from work. The share of walking and bicycle traffic is as high as 51.1%.

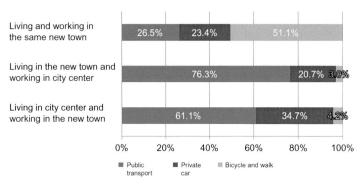

FIGURE 2.6 Stockholm commuting traffic share [5]. *From Robert Cervero. The Transit Metropolis [M]. USA: Island Press, 1998.*

FIGURE 2.7 Bicycle-only road on the main road of the city.

2.5 Walking and cycling traffic

While promoting the development of public transport, Stockholm also attaches great importance to the development of nonmotorized modes of transportation such as bicycles and pedestrian traffic.

In order to increase the proportion of bicycle travel, creating a good riding environment is a must. Stockholm has actively promoted the construction of bicycle roads. As shown in Fig. 2.7, on main roads, a bicycle-only road is constructed in parallel with the motor vehicle road. There are clear bicycle road signs on the road, and up and down lanes are separately set. Bicycle roads are separated from motorway and pedestrian roads, without interference from motor vehicles or pedestrians; independent lights for bicycles are

FIGURE 2.8   Special signal lights for bicycle roads at intersections.

FIGURE 2.9   An elegant, tree-lined bicycle-only road.

set up at intersections, as shown in Fig. 2.8. In some areas, bicycle roads and motor vehicle lanes are completely separate, with the bicycle road having an elegant environment and green trees, as shown in Fig. 2.9, with obvious bicycle signage. The construction of these road facilities ensures the safety, comfort, and speed of bicycle traffic.

In addition, Stockholm also has a public bicycle rental system, Stockholm City Bikes, which has more than 100 bicycle rental stations

FIGURE 2.10 Stockholm City public bicycle rental station. *From http://www.flickr.com/photos/question_everything/450953112/.*

throughout the city. People can choose to buy quarterly or 3-day cards to rent bicycles (Fig. 2.10).

Stockholm also gives careful consideration to walking safety. A pedestrian refugee island is usually set up in the middle of crosswalks, as shown in Fig. 2.11, providing a safe waiting space for pedestrians. In some old historical blocks or shopping streets, pedestrian-only roads are set up, as shown in Fig. 2.12. Some pedestrian roads and leisure seats are set up in waterfront areas, as shown in Fig. 2.13, creating a warm and comfortable walking environment for pedestrians.

2.6 Case study of a new town construction

2.6.1 Stockholm's first generation of new towns—Vallingby community as a representative example

The construction of new towns in Stockholm began in 1945. At the start, the concept of "ABC town" was proposed, namely, the construction of a new town of residence, employment, and service. Markelius's "half and a half" population planning principles guided the construction of Stockholm's first generation of new towns, and Vallingby is a typical representative example.

FIGURE 2.11   A safe island in the middle of the street to provide a safe space for pedestrians crossing the street.

FIGURE 2.12   Stockholm pedestrian-only shopping street.

Built in 1950–54, Vallingby community was the first new town built after World War II in Stockholm. It covers an area of 170 ha and is built on a mountain. It is located 13 km west of the city center and is connected to the city center by metro. As mentioned earlier, the designers of the first generation of new towns have two expectations: first, the new city should become a self-balancing community with balanced employment and

FIGURE 2.13 Stockholm waterfront walking path.

FIGURE 2.14 Aerial view of the Vallingby community. *From http://www.white.se/en/project/ 90-vallingby-city/slideshow.*

residence, reducing the cross-regional traffic; second, the new town and the central city should rely on rail transit.

Fig. 2.14 is the view of Vallingby community. It can be seen that the Vallingby community center business district is built around a subway station, surrounded by residential buildings. The business district has about 125 commercial retail stores, 7000 m^2 of community service spaces, more than

FIGURE 2.15   Shop street near Vallingby subway station.

30 restaurants and entertainment venues, dozens of retail suppliers, and more than 10,000 m² of office space. These commercial facilities have attracted a large number of people to the Vallingby community, which has boosted the vitality of the community and provided jobs that ensure job–housing balance. Fig. 2.15 shows a shopping street near Vallingby subway station.

Although Vallingby finally failed to fully achieve the goal of job–housing self-balance, through this first-generation new town construction, Stockholm planners gradually realized that it is difficult to achieve a complete balance in an individual community. Therefore in the second generation of new towns, all of the towns stand as a whole to achieve job–housing balance.

Vallingby community has a total population of 25,000 people, with the houses distributed over a 170-ha area around the commercial center. If you work in Vallingby, the walking time from home to work is generally less than 15 min. According to Stockholm statistics, about half of the travels in the new town are walking and cycling, with around 25% of travelers using public transportation, and only a quarter using cars. It can be seen that the construction of a new town with balanced employment and residence can greatly reduce the motor vehicle transportation and so reduce the pressure on urban traffic. On the other hand, more than 70% of cross-regional trips from or to the city center use public transportation [3], which shows that public transport is preferred for long-distance travels.

However, there are also some problems in the Vallingby community. The design of community buildings is relatively casual. Most are high-rise apartment buildings, as shown in Fig. 2.16. Most residents who live in the

FIGURE 2.16   High-rise residential apartment in Vallingby community.

community are low- or middle-income groups. There is a considerable amount of garbage (including cigarette waste) in the commercial area, and the community environment management is relatively poor.

2.6.2   Model of a sustainable community—Hammarby community

Hammarby community is located 6 km southeast of central Stockholm. The site of this community was originally the old industrial and port area. In the 1980s, Stockholm planned the Hammarby area as a modern new town on the waterfront. The community emphasizes ecological environment protection. The community began construction in 1997 and is expected to be completed in 2020. It had completed 75% of its construction by 2015, and has preliminarily formed a beautiful environment with waterways in the community, as shown in Fig. 2.17.

The Hammarby community had a population of 28,000 and a working population of 10,000 in 2015; the area is 204 ha (of which the land area is 171 ha). It has 11,500 sets of housing, with 33% for public rental housing, and 67% for commercial housing. Living houses generally have two bedrooms, primarily for family living.

The construction of energy-saving and environmental protection facilities in the Hammarby community has unique characteristics. At first, the community construction planning requires the regional impact on the environment to be reduced by 50%. The community has successfully transformed a harsh industrial and port land into a modern sustainable urban space. The Hammarby community also proposes the concept of "SymbioCity" (symbiotic city), which is a coordinated whole of landscape planning, water supply

FIGURE 2.17   The beautiful environment and waterways in Hammarby community.

and water quality assurance, waste utilization, building energy conservation, energy supply, and transportation. The Hammarby community also served as a demonstration project for Stockholm's bid to host the 2004 Olympic Games.

In order to create a sound ecological environment, the community has designed a rainwater-collecting system and waterfront landscape. The water system flows along each building so that all residents can enjoy the waterfront space, as shown in Fig. 2.18. Fig. 2.19 shows the fountains in the community and the green environment around the buildings. The Central Green Park, shown in Fig. 2.20, enhances the landscape quality of the entire community and provides a place for residents to communicate and rest.

The Hammarby community has a tram line connection to the central area of the city, as shown in Fig. 2.21. The tram line has been developed and constructed with the community construction process. The idea of constructing compliance into the plan and public transportation-oriented development has been implemented. At present, 75% of Hammarby community journeys are by public transport, with 12.5% relying on private cars, and 12.5% using bicycles and walking. There is also a ferry service from Stockholm city center to the Hammarby community. Energy efficiency is improved due to the relatively short water distance and large carrying capacity. The community has also introduced car-sharing and car rental services. Fig. 2.22 shows a vehicle that provides the service. Through ferry and car-sharing services, the community motor vehicle usage rate has been reduced by 40%. A public transport-led and green transportation concept has been implemented.

FIGURE 2.18 Each building in Hammarby community enjoys a waterfront space.

FIGURE 2.19 Fountains and green landscapes in Hammarby community.

The Hammarby community has consistently pursued the goal of reducing the negative impact on the environment by half during construction and operation. Biogas from the community facilitated through garbage collection is used for community heating and residential kitchen fuel. At the same time,

FIGURE 2.20 Waterfront landscape and walking path system in Hammarby community.

FIGURE 2.21 Tram line between Hammarby community and the city center.

eco-powered vehicles powered by biogas (Fig. 2.23) make full use of these resources. In the practice of trash classification, the community uses vacuum pumps to collect garbage. As shown in Fig. 2.24, each household only needs to classify the trash into different pipelines, and the community automatically uses the vacuum pipeline to collect the trash. In terms of water saving, the community has strengthened the construction of water-saving facilities, increased public awareness, and effectively reduced the per capita water

FIGURE 2.22   Vehicles for the vehicle-sharing and rental service in Hammarby community.

FIGURE 2.23   Stockholm biogas fuel bus.

consumption of the community from 200 to 150 L/day. It is quite difficult to achieve such low water consumption in Stockholm, a city surrounded by water. The per capita water consumption of cities in southern China, such as Shanghai and Guangzhou, is more than 200 L/day. Through the all-round promotion of community energy saving and environmental protection, the

FIGURE 2.24   Waste recycling vacuum facility.

per capita energy consumption of the Hammarby community has decreased by 30%−40% compared with other parts in Stockholm. Although there is still a gap to achieving the goal of halving this consumption, the progress so far has been quite significant.

The Hammarby community attaches great importance to public participation in planning and construction, and has set up a planning exhibition hall at the core of the community. As shown in Fig. 2.25, the model is used to display the layout of the community, and there is staff present to introduce the community to visiting residents, thus helping residents to understand and participate in the community planning and construction.

2.7   Summary of the characteristics and transportation experience in Stockholm

1. *Full implementation of the concept of job−housing balance*

In the process of the development of European cities, gradual differentiation of urban functions has generally appeared. The high value-added and competitive industries, such as business and service industries, have entered the central area of cities, and raised the land price in this

FIGURE 2.25 The planning exhibition hall in Hammarby community.

part of cities, causing residential areas to gradually move outward. The gradual differentiation of urban functions has led to huge amounts of commuting traffic between urban areas, which has brought enormous burdens to these cities.

The fundamental issue to improving urban transportation lies in a rational urban land use structure. Stockholm had already begun to practice this in the construction of new towns in the 1950s. Through job—housing balance in the community and new towns, people's cross-regional traffic has been greatly reduced.

2. *Give priority to public transportation*

Stockholm attaches great importance to improving the service level of public transport and promoting public transport priority. Stockholm relies on high-level public transport services and time-based congestion fee policies to increase the public transport share. Figs. 2.26 and 2.27 show the trams in Stockholm.

3. *Solidly promote bicycle traffic*

Through the construction of bicycle lanes and bicycle rental services, Stockholm has improved the cycling environment, and promoted bicycle traffic development.

4. *Pay attention to public participation in urban planning and construction*

In European cities, urban and community construction projects are the result of the participation of the government, planners, and the public. All projects will ultimately serve the residents and employees in the community. Therefore the building form and functions should be based on the needs of people who live or work in that community. Based on this

FIGURE 2.26   Stockholm tram at the station.

FIGURE 2.27   Stockholm tram.

principle, in today's European cities, the development of each community is basically generated by the continuous communication among government, planners, and the public. Public participation in urban planning and construction is a very valuable European experience. The key to achieving public participation is to build a platform for communication among the government, planners, and the public.

First, it is a must to build a platform displaying information to the public. The government has set up planning exhibition halls or exhibition rooms in Hammarby community in Sweden, Almere community in the Netherlands, and the Hafencity industrial transformation area in Hamburg, Germany. The planning drawings and planning models for the next development of the community are placed in the pavilion. Citizens can visit them, and make an appointment with the staff to have them explained. These venues have built a platform for people to learn about regional planning, so that the public can understand the ideas and methods of the planners and the government, which is the basis for communication.

Second, a platform is need for the exchange of opinions. There are usually a number of government representatives on duty in the planning pavilion. The public's opinions on the planning can be directly fed back to the staff on duty and submitted to the government and planners for reference. For major planning matters, government departments need to hold hearings to understand the suggestions and requirements from the people. In France, all new construction and renovation projects need to be recognized by the people around the project. In many cases, the most time-consuming work in European planning and construction is continuously seeking the opinions of the relevant public and obtaining their consent to the project construction.

Third, a platform is needed for public monitoring. The public has the right to choose their living or working environment. They can reflect their opinions and suggestions on the urban transformation and development projects through various channels, and fully monitor and constrain the project developers. Planning projects cannot be implemented without the recognition of the people in the surrounding areas of the project.

Public participation in European urban planning depends not only on good platforms, but also on the civic awareness of public participation. These experiences in Europe are well worth learning from.

5. *Highlight cultural protection and urban characteristics, protect characteristic buildings and blocks*

As an ancient city-state in northern Europe, the Swedish architectural style is greatly influenced by neoclassicism. A large number of Gothic buildings exist and form their own unique style. Leaning against the mountains, facing the sea, Stockholm has a variety of distinctive waterfront buildings. Today, although Stockholm has high-rise and modern buildings, there are still many Gothic buildings in the city, which highlight the characteristics of the Nordic city and reflect the city's history. As shown from Figs. 2.28−2.32, the old buildings and streetscapes of the country's romanticism in the early 20th century, such as the old waterfront buildings, the Stockholm City Hall, the heavy stone buildings, the slate pavements, etc., are well preserved. These buildings absorb the

FIGURE 2.28   Buildings along the lake.

FIGURE 2.29   Waterfront historic buildings.

characteristics of western and southern European architecture, and they blend in with local traditional decoration. The body is thick, the proportion is well-balanced, and the details are exquisite. All of them greatly enhance the collective beauty.

FIGURE 2.30   Stockholm City Hall.

FIGURE 2.31   Streetscape with distinctive architecture.

Stockholm has always attached great importance to the protection of architecture and culture. The city further demonstrates its deep historical heritage by retaining a number of historical buildings. European cities have put a lot of effort into the protection of historic buildings and urban

FIGURE 2.32   Stone building and pavement.

FIGURE 2.33   Stockholm waterfront leisure space.

features. The public is also very aware of the protection of ancient buildings, and people have actively support the protection programs. Each city retains its unique characteristics bringing a great deal of tourism to Europe. Fig. 2.33 shows the Stockholm waterfront leisure space, where people share these landscape resources with comfort, leisure, and pleasure. Fig. 2.34 shows a streetscape dotted with flowers and trees.

FIGURE 2.34 Green beautified roads and community environment.

6. *Energy-saving and environmentally friendly urban community construction*

   Stockholm's awareness of energy conservation and environmental protection is very strong. Many urban practices of energy conservation and environmental protection have been carried out by its communities. The practice of energy conservation and environmental protection is carried out jointly from community planning concepts, community construction and operation processes, high-tech applications, and public education guidance.

   Taking the Hammarby community in Stockholm as an example, planners first proposed energy conservation and environmental protection goals to reduce the environmental impact. After that, people adhered to this goal in the construction and operation of the community. Through comprehensive measures such as trash sorting, trash collection, water-saving facilities construction, and community water-saving publicity, the community has effectively reduced energy consumption and adverse impacts on the environment. The community has become a world model for energy-saving and environmentally friendly urban community.

References

[1]   Stadsledningskontoret. Stockholm facts & figures 2013. Stockholm, Sweden; 2013.
[2]   [United States] Lewis Mumford. History of urban development: origin, evolution and prospects [C]. Beijing: China Building Industry Press; 2005.

[3]   Cervero R. Sustainable new towns: Stockholm's rail-served satellites. Cities 1995;12 (1):41—51.

[4]   Xiao W. The planning and construction of the new post-war Stockholm city and its enlightenment. Huazhong Build 2008;9:164—70.

[5]   Cervero R. The transit metropolis. USA: Island Press; 1998.

[6]   Zuqi M, Suzhen F, Kai Y. A review of the traffic congestion charge policy in Stockholm inner city. Urban Issues 2011;8:2—8.

[7]   City of Stockholm traffic administration. Analysis of traffic in Stockholm with special focus on the effects of the congestion tax, 2005-2008, summary. Stockholm; 2009.

[8]   Storstockholms Lokaltrafik. SL Annual Report 2012. Edita Vastra Aros, Sweden; 2013.

[9]   Wikipedia, "Stockholm". <http://en.wikipedia.org/wiki/Stockholm>.

Further reading

Xiaoling L. Experience and enlightenment of urban planning and construction management in the three nordic countries. Urban Rural Constr 2012;12:84—6.

Chapter 3

Singapore, Singapore

Chapter Outline

3.1 Overview of the city

The Republic of Singapore (Singapore for short), located at the southernmost point of the Asian continent, is the fourth-largest financial center in the world after New York, London, and Hong Kong, and its port is one of the five busiest in the world. In 2012, Singapore covered an area of 715.8 km^2, with a total population of 5,312,400 and a population density of 7422 people/km^2. Instead of dividing into provinces and cities, Singapore is divided into five administrative regions, namely, Central Region, North Region, West Region, East Region, and North-East Region, in a manner consistent with urban planning.

3.2 Urban structure and land use

Singapore is small and lacks natural resources such as fresh water and minerals. It has been a long-term goal for Singaporeans to develop and utilize land

Eco-Cities and Green Transport. DOI: https://doi.org/10.1016/B978-0-12-821516-6.00003-5

efficiently and reclaim land as much as possible to increase its land area. According to the Singapore Urban Redevelopment Authority (URA), the total population of Singapore will increase to 6.5−6.9 million by 2030.

Urban land planning in Singapore is regularly formulated and implemented by the URA. Urban land development planning consists of two parts: the Concept Plan and the Master Plan. The Concept Plan is a strategic land use and transportation development plan to ensure that future land use patterns can meet the needs of population growth and economic development; the Master Plan is a more detailed and concrete implementation plan formulated after the Concept Plan has been determined, which specifically guides and regulates the development and use of urban land.

According to the 2010 Urban Master Plan, urban land planning data for Singapore in 2010 and 2030 are shown in Table 3.1. As can be seen from this table, with the population growth, there has been a housing, industrial, and commercial land increase of nearly 3%, while the road traffic infrastructure land has been controlled with a growth of 0.9%, and recreational and

TABLE 3.1 Comparison of land planning in Singapore in 2010 and 2030 [1].

Land type	2010 Area (km²)	Percentage	2030 Area (km²)	Percentage	Percentage variation
Housing	100	14.3	130	17.0	2.7
Industrial and commercial land	97	13.8	128	16.7	2.9
Natural reserve	57	8.1	72.5	9.5	1.4
Community, institutional, and entertainment land	54	7.7	55	7.2	− 0.5
Urban power and gas supplies	18.5	2.6	26	3.4	0.8
Impoundment	37	5.3	37	4.8	− 0.5
Road traffic infrastructure	83	11.8	97	12.7	0.9
Waterport, airport	22	3.1	44	5.7	2.6
National defense land	133	19.0	148	19.3	0.3
Other	100	14.3	28	3.7	− 9.6
Total	701.5	100	765.5	100	

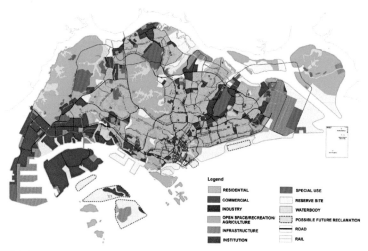

FIGURE 3.1   Singapore land planning map 2030 [1].

water storage land have been reduced by 0.5%. Therefore, continuing to control the use of cars and vigorously promoting green transportation are daunting tasks facing Singapore's urban land planning for the next 20 years.

Fig. 3.1 details the distribution of different types of land use in each planning area in Singapore in 2030. The light yellow indicates residential land, blue is commercial land, purple is industrial land, and green is open space or entertainment land. As can be seen from this figure, commercial land in Singapore is mainly concentrated in the south of the central area, with industrial land mainly concentrated in the south of the western area, although it is well dispersed, while residential land is scattered throughout the various planning areas, with open space or entertainment land mainly concentrated in the north and western areas, and a waterbody in each green park.

In 1971, Singapore formulated the first Concept Plan for urban land development, which identified the location of Changi airport and the prototype of the network structure of the urban rail transit system (Singapore Mass Rapid Transport, SMRT) and the highway system, laid the foundation for the planning of the public transport infrastructure in Singapore, and developed the Singapore future development model of "ring-shaped" land use: with the central aquifer as the center, and a number of high-density satellite cities being developed in a circular manner. Each satellite city has a commercial center, and a residential and light industry around the commercial center, which makes the satellite city a mixed area of commercial, residential, and light industries. Two adjacent satellite cities are separated by green spaces, parks, or open spaces to ensure a certain percentage of green space is retained. With the implementation of the "ring-shaped" plan, the central region of the southern coastline has gradually developed into an international economic and commercial center.

In 1991, Singapore revised its first-generation Concept Plan. On the basis of the original concept, it put forward a land use hierarchical decentralization strategy, to disperse the regional centers, subregional centers, and fringe centers at a hierarchical level across the island. The mixed commercial, residential, and industrial development reduced the average distance between residential areas and workplaces, achieving job–housing balance and alleviating urban congestion. The goal of the Concept Plan is to build Singapore into an island city with the functions of nature, beautification, and urban modernization, suitable for people's life, work, shopping, leisure, culture, and entertainment. After nearly 20 years of implementation, Tampines Regional Center and the Novina Fringe Regional Center are relatively successful cases in terms of job–housing balance and mixed land development.

In 2011, based on its forecast that the population would increase to 6.5–6.9 million by 2030, the URA proposed a future strategy for sustainable development in its Concept Plan, in order to maintain a high-quality living environment for Singaporeans, including the following main elements:

- Provide affordable housing subsidized by the government;
- Introduce green vegetation into living areas;
- Increase traffic mobility and network connectivity;
- Provide good employment opportunities and maintain a vibrant economy;
- Ensure future growth space and a good living environment.

At present, the Singapore government and related departments are working to implement new planning tasks under the framework of the future strategy for sustainable development.

3.3 Urban traffic system and traffic demand characteristics

As an economically developed island country, after nearly 50 years of development, Singapore has formed an urban transport system dominated by public transport systems (including subways, light rail, and bus) and a well-developed high-speed road system.

3.3.1 Rail transit system

The Singapore rail transit system comprises the subway (Singapore Mass Rapid Transport, MRT) and the light rail (LRT), which share about 2.65 million passengers per day, accounting for about 37.3% [2] of Singapore's total daily passenger transport. Fig. 3.2 is a map of the MRT and LRT lines. In 2012, the total length of the MRT lines was 148.9 km, with 99 stations; the total length of the light rail lines is 28.8 km, with 34 stations [3].

MRT & LRT System map

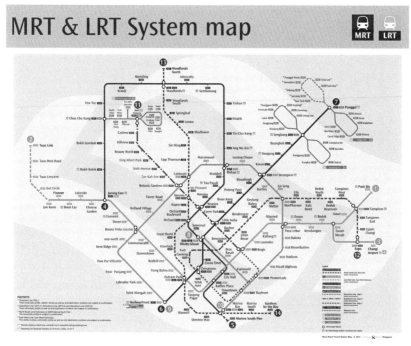

FIGURE 3.2   Rail transit system map of Singapore in 2012.

Singapore Mass Rapid Transport (MRT) consists of five main lines, namely, the North South Line (red), East West Line (green), North East Line (purple), Circle Line (orange), and Downtown Line (blue; under construction). Singapore MRT hub adopts the mode of comprehensive development transportation and land development, through the combination of transportation planning and land use, urban residents can not only complete seamless transfer between different lines and modes of transportation at the hub, but also carry out daily activities such as shopping, dining, and leisure. The Transit-Oriented Development (TOD) model of comprehensive development of transportation and land integration has been excellently realized.

As can be seen from the MRT network map, Singapore rail transit network layout has a number of small loops that are not available in other cities. It is this small loop in large residential areas that supports the high-intensity development of a bedroom community in Singapore. Generally speaking, urban planning should avoid sleeping city construction. However, it is difficult to build more houses and realize mixed land use in the central area of Singapore. With this special background, Singapore has successfully solved the traffic problems caused by the superintensity land development by adopting branch rail transit to distribute the powerful traffic demand of large-scale residential areas, and it is also a problem-solving idea which is closely combined with Singapore's actual conditions.

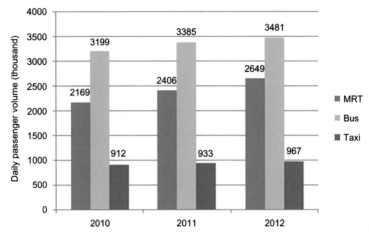

FIGURE 3.3 Daily passenger volume changes of the three public modes (MRT, bus, and taxi) in 2010—12 [2].

Fig. 3.3 shows the daily passenger volume change of MRT, bus, and taxi from 2010 to 2012, showing that Singapore's daily passenger volume was increasing slowly.

3.3.2 Bus system

In addition to a sound rail transit system, Singapore also has a bus network that extends all over the island. Singapore buses are operated by two major companies, the SMRT and Singapore Bus Service (SBS). There are 344 service lines and more than 4600 bus stops on the island. According to the statistical data from 2012, bus travel accounts for about 50% of total public transport [2], which is even higher than that of the rail transit system.

In order to further improve the service quality of the public transport system, the Singapore government began to modernize the management of the public transport system in 2010, aiming at enhancing the service quality of the public transport system and reducing the use of private vehicles. The measures include two aspects: establishing a more effective bus priority system and intelligent bus toll and inquiry system. The bus priority system includes setting up dedicated bus lanes, bus priority signal lights, and compulsory signs for car concessions at bus stops.

3.3.3 Vehicle ownership and road network

By 2012, Singapore had 969,910 motor vehicles, including 535,233 private vehicles. The Singapore road network consists of nine main expressways, arterial roads, collector roads, and local access roads, with average daily

TABLE 3.2 Total length of roads by functional class in the Singapore road network (2012) [3].

Road type	Length (km)
Expressways	161
Arterial roads	652
Collector roads	561
Local access roads	2051

TABLE 3.3 Average daily passenger-journeys and average journey distance in Singapore [3].

	2004	2012
Average daily passenger-journeys (10,000 trips)	310	413
Average journey distance (km)	9.4	9.7

traffic volume reaching 300,000. Table 3.2 shows the length of all functional classes of roads in 2012, of which the total length of expressways is 161 km.

The Singapore road system is equipped with a comprehensive intelligent transportation sensing and detection system. By integrating loop detectors, video detectors, and floating car data, the Singapore Land Transport Authority (LTA) provides real-time road network flow and speed information (within 5 min). In addition, the Singapore government has adopted various intelligent transportation technologies to reduce traffic congestion and improve service quality. Among these, the most notable achievement is electronic road pricing (ERP), which adjusts the spatial and temporal distribution of traffic flow through dynamic toll collection.

3.3.4 Travel characteristics of residents

According to the data from the household interview travel survey 2012 (HITS 2012), from 2004 to 2012, Singapore traffic demand increased by almost 13%. As shown in Table 3.3, average daily passenger-journeys increased by 33.2% during the 8-year period, however, the average journey distance barely increased. This indicates that Singapore has achieved good results in promoting mixed land development, job—residence balance, and reducing resident travel distance.

TABLE 3.4 Share of different modes of transport in Singapore urban areas [4].

Mode of transport	Buses	MRT/ LRT	Taxis	Private vehicles	Total public transport
Share	28.8%	20.7%	7.2%	43.2%	49.5%

HITS 2012 also shows that the share rate for private vehicles increased by 29% from 2004 to 2008, while the share rate of public transport increased by only 13% during the same period. However, from 2008 to 2012, the share rate of private vehicles slowed significantly to 9% from 2008 to 2012, compared to the previous 4 years, while the share rate of public transport increased by 14% in the same period. In the peak period of 2012, the share rate of public transport reached 63%. It can be seen that through the continuous improvement of the public transport system and modern traffic management measures such as ERP, the Singapore government has effectively controlled the private vehicle use rate, thus ensuring the service level of the whole urban transportation system.

Table 3.4 lists the shares of different modes of transport in 2012. Among these, the private vehicle share is 43.2%, the total public transport (including bus and MRT/LRT) share is 49.5%, and the taxi share is 7.2%.

3.4   The concept and measures of the green sustainable transportation plan

Although Singapore already has a very well-developed transportation system and road congestion has been controlled to an extent, due to the limited land resources, 12% of the land area has been used for road transportation-related infrastructure construction. In the next 20 years, the Singapore government will no longer be able to use more land for road transportation infrastructure. Therefore the government is focused on building a green sustainable transportation system. A green sustainable transportation plan has been proposed, which mainly includes strengthening the public transportation system, and improving the energy utilization rate and the utilization of clean energy.

Strengthening the public transportation system includes: increasing the capacity of the existing rail network, a plan to double the total length of rail lines by 2020, implementing public transportation priority measures, and realizing a seamless transfer system between rail transit and bus, with a planned public transportation share of 70% in the peak period by 2020. In terms of transportation energy, the government will continue to control the growth of cars aiming to control the annual growth of cars at 1.5%. Through further improving the ERP system it is also planned to regulate and suppress the use rate of motor vehicles in peak hours.

At present, Singapore plans to implement ERP-II system in the future. The system will be upgraded according to the existing ERP system and realize dynamic charging according to the distance traveled and area, in order to reduce traffic congestion more effectively. The use of clean energy is another important aspect of the sustainable development mode. The Singapore government plans to realize the concept of green transportation from two aspects. One is to strengthen the detection and control of vehicle exhaust emissions. The other is to build more nonmotorized lanes to encourage travelers to use green transportation, such as bicycles and walking. In order to realize this green sustainable transportation plan, the Singapore government has formulated and implemented three specific plans: the Compulsory Bus Priority Plan, the ERP system, and the National Bicycle Plan.

3.4.1 Compulsory Bus Priority Plan

The Singapore LTA began to implement the "Compulsory Bus Priority Plan" in 2008 to improve the speed and reliability of buses. That is, when an ordinary car approaches a bus stop, the driver first sees a sign indicating to give way to buses, as shown in Fig. 3.4. At this time, the driver of an ordinary car needs to slow down, pay attention to whether there is a bus exiting the station, and if a bus is exiting, the driver of the ordinary car must stop or change lane before the line of giving way, so as not to affect the bus exit. If found to be in breach of this, ordinary car drivers will be fined S$130. The Compulsory Bus Priority Plan aims to greatly reduce the bus exit time, thus

FIGURE 3.4 Compulsory bus priority sign in front of a bus stop.

FIGURE 3.5 Cars give way to buses at bus stops.

shortening the bus travel time and improving their reliability and service quality. By February 2013, 218 bus stops were covered by this plan, and Singapore will add 150 more bus stops in this plan in the next 2 years. Fig. 3.5 shows that cars passing through bus stops give way to avoid affecting bus priority exiting.

3.4.2 Electronic road pricing system (electronic road pricing)

The predecessor of the ERP system was the regional license system, which had been implemented since 1975. This system charged a uniform congestion fee for all vehicles entering the Central Business District (CBD) area. Since 1998, this system has been replaced by the modern ERP system, which uses a more scientific, reasonable, and comprehensive congestion charging system. Singapore congestion pricing is based on Radio Frequency Identification (RFID) technology.

Since 2008, road detection and toll collection devices (ERP system) have been installed in the designated urban central area, as shown in Fig. 3.6. At the same time, traffic laws stipulate that if a motor vehicle needs to use toll roads, it must install a S$150 (about 750 RMB) in-vehicle unit to pay for congestion. The ERP system detects the 85-percentile speed of motor vehicles on the road as a criterion for deciding whether to charge congestion fees or not. When the speed on a highway is lower than 45 km/h, and the speed on an urban main road is lower than 20 km/h, the ERP system starts congestion charging. The ERP system is not only helpful in controlling traffic flow at peak times, but also can balance the temporal and spatial distribution of traffic flow in the network. It can improve the overall performance of the

FIGURE 3.6   Singapore urban road network ERP system.

network by encouraging vehicles to move to a route with less traffic flow density and higher speed.

3.4.3   National Cycling Plan

The National Cycling Plan, a representative project of the Singapore green sustainable transport plan, has been implemented since July 2010 with the aim of increasing the share of bicycles for short-distance journeys. As a successful case, Tampines Town has built an excellent bicycle-dedicated road network and achieved seamless connection with MTR and bus hubs. In the next 2 years, the Singapore government plans to build dedicated bicycle lanes in seven towns (Tampini, Yishun, Taman Jurong, Sembawang, Pasir Ris, Changyi Simei, and Bedok) and Marina Bay, and connect bicycle lanes to major transport hubs (such as MTR stations and bus transfer stations). There are now 440km of cycling paths in Singapore. It is planned to expand the cycling path network to 750km by 2025, and triple cycling network by 2030.

3.5   Garden city construction

Singapore has a reputation as a "garden city." Almost half of the city is covered by green vegetation. With the population growth in recent years, the Singapore government has integrated land use, environmental construction, and transportation planning, and put forward the goal of building a modern city of "garden and water" in the future. The proposal includes the construction of national parks and natural leisure areas, vigorous advocacy of "highrise vegetation," the development and utilization of reservoirs, and enhancement of biodiversity. It has established new goals and implementation strategies for building further garden cities, and has achieved positive initial results.

3.5.1 National park construction

Singapore plans to build another 900 ha of national parks on the basis of existing parks by 2030, reaching 4200 ha by 2020, and achieving an average of 0.8 ha per thousand people by 2030. In order to combine leisure and recreation spaces with green vegetation, the construction of new national parks will become the main mode of development in Singapore in the future.

Garden By the Bay, built in 2011, is a representative achievement of the Singapore national park construction plan. Situated on the southeast side of Marina Bay Sands, this 2-km^2 park is built entirely on reclaimed land. The shellfish-shaped indoor botanical garden (Fig. 3.7) in the park is well coordinated with the coastal geographic environment. Creative manmade towering trees stand on the ground, exposing metal materials. Singaporeans have begun to cultivate green vegetation at the roots and facades of manmade trees, as shown in Fig. 3.8. These will take on a completely natural appearance after several years and include green lawns for people to enjoy (Fig. 3.9). Garden By the Bay, which is integrated with Marina Bay Sands and shopping mall, is a good place for citizens to spend their holidays, shopping, catering, recreation, and exercise. In the next 20 years, Singapore will continue to build national parks to make Singapore a real garden city.

FIGURE 3.7 Singapore Garden By the Bay.

FIGURE 3.8 Cultivating green vegetation at the roots and facades of manmade trees.

FIGURE 3.9 Green lawn for the leisure and recreation of citizens and tourists.

3.5.2 Promoting vegetation in high-rise buildings and walls to create a three-dimensional green space

Singapore is an international metropolis with scarce land resources and dense population. Most of its urban land is occupied by high-rise buildings and transportation infrastructure. In order to increase the greening of the urban central area and high-density residential areas and to maintain the good ecological environment of the city, Singapore has put forward the concept of "skyrise greenery" and vigorously advocated greening of the roofs and elevations of buildings. "Skyrise greenery" can also be divided into rooftop greenery and vertical greenery, as shown in Fig. 3.10A and B, respectively.

As shown in Fig. 3.11, the famous Marina Bay Sands and shopping mall, whose architectural design is like a giant ship sailing in the Pacific Ocean, has 55 floors. It is a landmark building in Singapore. On the roof of the building, a green garden was built at a height of 198 m. It can be regarded as

FIGURE 3.10 Two types of "skyrise greenery" [5]: (A) rooftop greenery; (B) vertical greenery.

FIGURE 3.11   High-rise greening: Marina Bay Sands Sky Park.

FIGURE 3.12   Wall surface greenery.

a representative of Singapore in promoting high-rise greening, and is a very innovative architectural design.

At present, the concept of "skyrise greenery" has been adopted by many buildings in Singapore. As shown in Figs. 3.12 and 3.13, it has achieved good results in reducing the "heat-island effect," absorbing noise, improving the landscape, and avoiding the phenomenon of an "urban desert."

FIGURE 3.13 Vertical greenery of buildings.

3.5.3 Building a cozy walking system

In view of the local climate characteristics, intense sunshine, and muggy and rainy weather, the government and relevant departments attach great importance to the construction of corridors, sunshade roofs, and other facilities between public facilities, such as buildings, bus stations, and subway stations, so as to create a continuous and comfortable walking system for citizens. Many commuters barely need to walk in the open air when commuting, and often do not need to take umbrellas during rainy days. The government and relevant departments pay attention to the landscape design of the pedestrian system to provide comfortable walking conditions for pedestrians.

Fig. 3.14 shows a corridor between two buildings. Pedestrians can reach another building through the air corridor between two buildings, thus avoiding outdoor sunshine and heat, and cross roads using bridges which separate pedestrians and motor vehicles. Fig. 3.15 shows a building corridor, half in the open air, providing convenience for pedestrians to shelter from the sun and rain. In the high-rise central business district, there are many such air corridors and outdoor corridors constructed on the buildings themselves. Fig. 3.16 shows a bus stop located at the bottom of a building. The upper building makes a wide sunshade roof. Travelers can enter office buildings or shops after alighting from a bus without being affected by the sun or rain. Figs. 3.17 and 3.18 are open sidewalks with shade roofs that are exposed to nature on both sides. Pedestrians thus make contact with nature while avoiding sunshine and rain.

In Singapore, there are various designs of pedestrian walkways. In addition to considering the physical needs of shelter from sunshine and rain, there

FIGURE 3.22   Walkway with trees outside the building.

FIGURE 3.23   Boardwalk along the water.

1. Integration of transportation and land use planning and implementation of integrated development. Singapore integrates residential, commercial, and industrial facilities with the rail transit system, building a compact

FIGURE 3.24 Boardwalk far over the water.

FIGURE 3.25 Fans in the park.

city, emphasizing the job–housing balance, with schools and offices located nearby, and shortening the travel distances of residents. Due to the lack of land resources in Singapore, a commercial development mode in the central area has been selected. Residential areas are also developed in high floor area ratios, high density, and on a large scale. This form of development is not advisable and usually leads to unbearable traffic demand, however it is required by the extremely scarce land conditions in Singapore. In order to achieve high-intensity land development while also solving its traffic problem, Singapore has achieved the perfect integration of a transportation system and land use development, and adopted branch rail transit to distribute the significant traffic demand of large-scale residential areas. This is an innovative integration of land use and transportation system (the spatial characteristics of rail transit lines in Singapore are unique). Singapore's three-dimensional pedestrian system has very smooth coordination with the urban design stage. For example, when setting up an air corridor between two buildings, the government coordinates the relevant buildings to determine the location of the air corridor and the division of construction work, which is then incorporated into the normal development process with strict planning control, to ensure the realization of the walkway and architecture integration in the planning, design, and construction stages.

2. Implementation of public transport priority policy has resulted in the government taking rail transit as the transport backbone, expanding bus coverage, and ensuring a seamless connection among the various modes of transport through bus hubs, in order to improve the share of bus and rail transit. In the development of large residential areas, simultaneous construction of public transport provides options and guarantees for citizens to give priority to the choice of public transport modes.

3. According to its limited land resources, in the interests of the whole country, Singapore has steadfastly implemented traffic demand management, restrained the ownership and use of private vehicles, and implemented congestion charging and safety management for vehicles by using advanced technology of intelligent systems, achieving remarkable results.

4. The Singapore pedestrian system is quite out of the ordinary. In urban built-up areas in Singapore, pedestrian space is a three-dimensional pedestrian system connected by air corridors, public space plaza platforms, large shopping malls, and overpasses. Due to the hot and rainy climate, canopies are often in place. This type of design is a unique choice in cities with a high density of buildings and scarce land.

5. Despite its limited land resources, Singapore attaches great importance to the establishment of an ecological environment and green space. It has not only set up large-scale parks in the Marina Bay area, but also has shady and flourishing airport roads. Singapore pays attention to high-rise vegetation, three-dimensional greenery, and has created a livable city with ecological environment protection and beautiful scenery. The Singapore design philosophy is to move from a garden in the city to a city in the garden.

References

[1] URA (Urban Redevelopment Authority) website. Singapore master plan 2008, <http://www.ura.gov.sg/uol/master-plan.aspx?P1 = View-Master-Plan&p2 = Master-Plan-2008>.

[2] LTA (Land Transport Authority) website. Publications and research, <http://www.lta.gov.sg/content/dam/ltaweb/corp/Publications Research/files/Facts and Figures/PT%20 Ridership %20(2012)>.

[3] LTA website, Singapore Land Transport: statistics in brief 2004−2013, <http://www.lta.gov.sg/content/ltaweb/en/publications-and-research.html>.

[4] LTA (Land Transport Authority) website. Household interview travel survey 2012, Annex A, <http://app.lta.gov.sg/data/apps/news/press/2013/07-10-2013%20Annex_A_Fig1_Travel_demand_Fig2_PT_mode share.pdf>.

[5] Skyrise Greenery website, <http://www.skyrisegreenery.com/index.php/home/skyrise/rooftop_greenery_8211_the_horizontal_dimension/>, <http://www.skyrisegreenery.com/index.php/home/skyrise/vertical_greenery_8211_the_vertical_dimension_dimension_dimension/>, <http://www.skyrisegreenery.com/index.php/home/skyrise/vertical_greenery_8211_the_vertical_dimension/>.

Chapter 4

Curitiba, Brazil

Chapter Outline

4.1 Overview of the city

Curitiba is located in the highlands of Serra do Mar Mountain in southern Brazil at an altitude of 900 m. Founded in 1654, it was originally a gold mining area and became the capital of Paraná in 1854. With an urban area of 430.9 km^2, it is the seventh largest city in Brazil. Curitiba city has a population of 1.9 million, a population density of 4418 people/km^2, and 400 motor vehicles per thousand people, as shown in Table 4.1. As early as the 1980s, Curitiba was hailed as the cleanest city in Brazil. In 1990, Curitiba was awarded the highest environmental award by the United Nations, and in 2001 it was ranked as having the highest living index in Brazil by the United Nations. Curitiba, together with Vancouver, Paris, Rome, and Sydney, was named "the most suitable city for human settlements" by the United Nations in the first batch.

4.2 Characteristics of the city

4.2.1 Green city

Curitiba is one of the greenest cities in the world, and is known as an ecological city. There are 30 large parks and forest parks, and as many as 200 street parks and green spaces. The per capita green area of Curitiba is 51 m^2, more than three times the United Nations standard of 16 m^2/capita. Its urban greening adopts a method of combining artificial planting with the natural ecology. Natural grassland can be grazed without fear of being walked on by

Eco-Cities and Green Transport. DOI: https://doi.org/10.1016/B978-0-12-821516-6.00004-7

TABLE 4.1 Basic data list for Curitiba.

Urban area	430.9 km^2
Urban population	1.9 million
Population density	4418 people/km^2
Car ownership per thousand people	400 vehicles per thousand people
Green space area per capita	51 m^2
Annual income per capita	2500 US dollars

people and animals. Artificial grassland uses native grass species that have strong vitality and adaptability. The grasslands are all directly connected by roads and walkways, and are skillfully integrated with urban buildings. In addition, great importance is attached to the reuse of waste mines and garbage dumps by transforming them into ecological parks, which greatly improves the landscape environment of the city and creates a good living environment for residents.

4.2.2   Bus city

Curitiba is internationally recognized as a model city for public transport. Reflecting its unique situation, the city has built an integrated public transport system at reasonable cost. This system has been planned and constructed since 1972. The first north–south axis, with a total length of 20 km, was completed in 1973, and the line was put into operation in 1974. The terminals at both ends of the line are connected to the Bus Rapid Transit (BRT) through a 45-km bus barge line, with an average daily passenger flow of about 45,000. In 1978, the southeast axis, about 9 km, was built to form a new development axis. The concept of an integrated public transport network was created in 1979.

At that time, the service through the urban area was relatively weak, and so the interregional transport services was put on the agenda. The initial interregional service was a 44-km ring line that contacted three BRT axes through the intermediate hubs. In 1980, there were nine intermediate and terminal stations where passengers could transfer between express lines, barge lines, and interregional lines, with an average daily passenger flow of more than 200,000. In 1991, all five radial axes were completed, with two connecting lines being added to the south to connect two axes. The total length of the bus lanes reached 72 km. From 1974 to 1994, the average annual passenger flow growth rate in the public transport system was 15%, which was four times the population growth rate. The proportion of commuting trips using public transport increased from 8% to 70%. Most of this growth was attributed to the express lines that opened in the 1970s, the interregional and district lines that opened in the 1980s, and the direct lines that opened in the 1990s.

With the expansion of the BRT line network, the number of hub stations has also increased constantly. There are currently 33 hub stations. The bus network connected to the hub is expanding, and the area covered by the integrated public transport system is also expanding. At present, the Curitiba bus system consists of more than 390 routes and almost 2200 buses, with a daily passenger volume of more than 2 million, covering more than 1100 km of roadways in Curitiba. After more than more than 30 years of continuous improvement, a Metrobus (BRT) system with sound functional hierarchy and clear division of labor has finally been formed. The system is currently composed of five lines with different service functions. The functions of each line and operating vehicle characteristics are described in Table 4.2.

TABLE 4.2 Functional hierarchy of the Metrobus system.

Line name	Function and operation characteristics	Vehicle characteristics
Express line	Connecting the downtown area with the integrated hub station, it runs on a dedicated road, and the entry and exit stations are realized through a cylinder station	Red dual-articulated vehicle operation
Barge line	Connecting the integrated hub station and its vicinity	Orange vehicle operation
Interregional line	Connecting several surrounding urban areas and integrated hub stations, not reaching the central area	Green vehicle operation, long single locomotive and articulated vehicle
Direct line	As a supplementary line to express lines and interregional lines, the average distance between stations is 3 km, and the entry and exit stations pass through cylinder stations	Silver single locomotive operation
Main line	Connecting the integrated station with the central area, using general roads	Yellow vehicle operation, with standard, long, and articulated vehicles
Conventional line	Connecting the surrounding cities with the central area, there is no integration with other public transport	Yellow vehicle operation, with standard and long vehicles

The layout of dedicated bus lane in Curitiba is closely integrated with the overall planning of the developing city. Five radiation lines of the rapid transit system constitute the main development axes of the city. The dense residential and commercial areas are concentrated near the bus stops and are developed along the bus line. The floor area ratio of areas along the four of five radiation lines are 6, while that of areas along other bus lines are 4. The farther away from the bus line, the lower the floor area ratio is. The integration of land use, road system, and rapid transit has been realized.

In terms of the road system, the integrated road hierarchy system consists of three levels: urban axis, urban main road, and urban branch road. Among these, the urban axis is the main traffic corridor for the city, the urban main road is the axis to distribute urban traffic, and the urban branch serves all kinds of land use. Curitiba's axial road system adopts a "trinary" structural concept: each axis is composed of three parallel extended arterial roads. One of the arterial roads leads to the city center and the other departs from the city center. The two one-way road spaces are designed for vehicles entering and leaving the city center. They connect the city center of Curitiba with other major areas and surrounding areas of the city. They play the role of express lines and allow for easy transfer to express lines. Motor vehicles and silver buses run on them. The third road is the central road between two one-way roads. Roads are separated by standard urban blocks. Three parallel lanes are compactly arranged on the central road, with the BRT system in the middle, providing fast public transport services, and a pair of one-way fast arterial systems on both sides, providing fast lanes for motor vehicles. The BRT lanes are separated from the one-way motor lane on both sides by green belts. In the integrated public transport network, there are large bus stations at the junctions of the BRT and cycle lines. Large transit bus stations provide separated boarding and alighting platforms for different lines, and connect these platforms using underground passages, so that passengers can transfer between the central roads. At the end of the structural axis, large terminal bus stations can satisfy passengers' transfer needs, generally in the area of $1500-1800 \, \text{m}^2$. In order to make the Metrobus system play a better role in a large-capacity transportation system, 266 transparent tubular plastic bus platforms (Fig. 4.1) with $500-1000 \, \text{m}$ spacing were designed. These platforms are on the same level as the buses (800 mm off the ground). After the bus arrives, the doors are opened automatically and the elevation-free pedals covering the gap between the bus floor and the platform are unfolded. Passengers get on and off the bus conveniently (see Fig. 4.2). As in the subway, passengers can buy tickets from the conductor in advance when waiting on the platform, and then transfer freely to the system without having to buy additional tickets. Passengers get on and off at the same speed as the subway, in only $15-19$ seconds. The efficiency of the BRT is very high, with about 75% of commuters choosing to take the bus, and the number of trips per capita is 350 per year.

FIGURE 4.1   Curitiba tubular bus platform.

FIGURE 4.2   BRT vehicle and tubular bus platform.

Therefore, although the city has a high motor vehicle ownership rate, with an average of 400 motor vehicles per 1000 people, there is no traffic congestion. The journey time from the city center to any destination within the city is less than 1 hour, and the overall traffic condition of the city has reached a high level. In addition, due to the developed bus system, Curitiba's transportation fuel consumption is 25% of that of similarly sized cities. Compared with other cities in Brazil, Curitiba saves about 7 million gallons of fuel a year, reducing vehicle fuel consumption by about 30%, and urban air pollution is much lower than that of other similarly sized cities.

4.3 Summary of successful experiences in Curitiba

4.3.1 Government guides scientifically and solves problems systematically

Before the 1970s, like most Brazilian cities, Curitiba faced serious social and environmental problems including population congestion, poverty, unemployment, environmental pollution, etc. However, since the 1990s, Curitiba has become famous globally as an ecological and livable city. In this transformation process, the scientific management and guidance from the government have played a major role. Among these, Jaime Lerner (December 17, 1937), who served as mayor three times (1971–75, 1979–84, 1989–92), has played an important role in supporting Curitiba to achieve substantial leaps in sustainable development. When he became mayor, he adhered to the new concept of sustainable development and supplemented by a small amount of government investment, inspired the public by describing and implementing the systematic development strategies of "multidepartment management," "creativity and efficiency," "people-oriented," "respect for citizens, regarding citizens as owners and participants of all public assets and services," "green city," etc. With the imagination and enthusiasm of the public, in only one generation, under the premise of protecting and developing the ecological environment, it has fundamentally improved the urban landscape and greatly improved the quality of life of residents. It has also gained the consensus and support of the people and successive mayors, which has made Curitiba move toward the goal of sustainable development. In 1990, Curitiba became the only city in the developing world named by the United Nations as "the most suitable city for human settlement" (the other four cities being Vancouver, Paris, Rome, and Sydney). It is also one of the greenest cities in the world, and is known as the ecological city of Brazil. After more than 30 years of miraculous sustainable development, the citizens of Curitiba generally believe that they live in one of the best cities in the world, putting to shame many cities in developed countries. In the process of creating the miracle of substantial leaps forward in the sustainable development of Curitiba, it was faced many difficulties. Mayor Jamie Lerner has shown remarkable systematic thinking. He has always emphasized that "we cannot solve one problem, but cause more problems. We should try to link all problems into one problem, treat them systematically, and solve them by means of comprehensive planning." His famous saying "City is not a problem, city is a solution" is widely known throughout the world.

4.3.2 Policies for classified garbage recycling

In order to control environmental pollution and cope with the increasing shortage of landfill capacity, in 1989, the municipal government of Curitiba put forward the slogan "garbage is not waste, garbage is also resources," and launched a campaign called "Garbage that is Not Garbage." Mobilizing families in the city to separate recyclable materials from garbage, the municipal government allocated special funds to implement the "Green Change" program in communities throughout the city, guiding citizens to collect paper, metals, plastics, glass, oil, and other garbage from their homes and send them to the designated exchange sites. These are exchanged for seasonal surplus products from local farmers such as tomatoes, potatoes, garlic, and bananas. The exchange standard is that 4 kg of garbage can be exchanged for 1 kg of food, in addition to bus tickets, stationery, and toys. The implementation of this program not only protects the urban environment, but also meets the daily needs of citizens, and increases farmers' income. It can be said that it kills three birds with one stone. Currently, there are 95 exchange sites in Curitiba, which are located in various communities around the city. The sites are open every 15 days, from 9 a.m. to 11 a.m. According to the figures provided by the municipal government of Curitiba, the classified garbage recycling rate of Curitiba has reached 70%, with 85% of the citizens of Curitiba having participated in the classified garbage recycling scheme since the implementation of the "Green Change" program 24 years ago. The recycled garbage is classified and sold to special recycling companies for reprocessing, which effectively realizes the reduction and resource utilization of closed-circuit recycling garbage. So far, the amount of waste paper recycled has been equivalent to 1200 trees/day. A local landfill was already nearly saturated 15 years ago. However, due to the implementation of the "Green Change" program it was used for another 15 years. Curitiba classified garbage recycling is not limited to this—under the guidance of the government, awareness of the treatment of toxic and harmful garbage is also increasing. Toxic garbage includes four kinds of waste, including electrical appliances and medical waste. In order to prevent them from polluting the environment, Curitiba has selected 23 bus stations in the urban area, which are used from Monday to Saturday, cars arrive at the 23 bus stations to collect the waste free of charge every day. Currently, the amount of toxic and harmful garbage collected accounts for 22% of the total amount of garbage recycling. These items are sent to special treatment plants for classification, which effectively reduces any environmental pollution.

In addition to positive incentives, the municipal government has also introduced appropriate penalties to ensure the smooth development of classified garbage recycling: if citizens discard of waste incorrectly, they can be subject to a minimum fine of 400 reals (about 92.5 US dollars) and a maximum fine of 65,000 reals (about 15000 US dollars).

Waste treatment is a major challenge facing modern metropolises. Curitiba has pioneered a new path in this respect. According to its own national conditions, Curitiba has developed a low-cost, high-efficiency, and sustainable development path, which provides useful experience for the urbanization construction and circular economy development in other countries, including China.

4.3.3 Successful practice of the bus rapid transit system

Currently, the public transport system in Curitiba is one of the most practical and best urban transport systems in the world, with a reasonable layout, scientific diversion, and each line having its own responsibilities. It provides a low-cost, fast speed, and punctual solution for densely populated cities with a short construction-period, which is worth learning for implementation elsewhere.

4.3.3.1 System characteristics of the Metrobus

Among the existing modes of public transport, metro, light rail, monorail, new transportation system, streetcar, and traditional electric bus have their advantages and disadvantages in terms of transport capacity, scope of application, cost, operation speed, and punctuality. Therefore they are suitable for different types of cities and different levels of social and economic development. At the same time, it is possible that mutual integration could further improve public transport service efficiency.

The Metrobus system is a new type of urban public transport system developed with this background. It has been successfully practiced in Curitiba. In order to make the system successful, on the one hand, it strengthens the system itself, that is, high-capacity vehicles, getting on and off, and selling tickets quickly and conveniently; on the other hand, it fully considers the traffic management countermeasures and modernization of traffic management, that is, the exclusive right of roads and the realization of real bus priority at intersections, as shown in Fig. 4.3.

Overall, the Metrobus system includes seven core parts, namely, modern public transport, exclusive right of roads, platform-level boarding, station ticketing, priority at intersection, passenger information, and fleet management. The effective combination of these seven parts gives the Metrobus system the following characteristics.

4.3.3.1.1 Large capacity

The unique large-capacity bus (see Fig. 4.4) of the Metrobus system increases its passenger-carrying rate. Meanwhile, exclusive right of roads and priority at intersections increases the speed of the bus system. Therefore, the single direction hourly section flow of BRT system is large, and it can reach capacity close to that of the light rail system under the unique land use pattern in Curitiba.

FIGURE 4.3   Curitiba BRT priority.

FIGURE 4.4   BRT large-capacity vehicles.

4.3.3.1.2   Low cost

The Metrobus (Curitiba BRT) system uses roads. It does not need to introduce dedicated rail vehicles, and only needs to reform existing roads with no need to build tracks. The construction work is limited, so the initial cost of the system is low and the construction speed is fast.

4.3.3.1.3 Good flexibility

The Metrobus system does not use tracks, and so a complete dedicated bus lane network is not needed. In some parts, High-Occupancy Vehicle (HOV) lanes or even ordinary roads can be used and shared with other modes of transport. Therefore the line network can be implemented in stages, with priority at intersections, passenger information systems, and other technologies also being introduced gradually. At the same time, using roads ensures that the line can be adjusted or changed more easily, even when the attracted traffic flow reaches the upper limit of the system, as a rail transit system with higher capacity can be built.

4.3.3.1.4 Fast speed and high reliability

The Metrobus system uses dedicated bus lanes and has priority at intersections (Fig. 4.3), therefore it is less disturbed by other modes of transportation, has high travel speed, and it is easy to meet planned schedules. In addition, the platform-level boarding system and the out-of-car ticketing system reduce the waiting time, shorten the travel time, and increase the average speed.

4.3.3.1.5 User-friendly

The new type of bus is spacious, comfortable, has less noise and vibration, and is more comfortable. The use of a platform-level boarding system makes it easier for passengers to board, especially passengers carrying parcels and those with a lack of mobility. In addition, the use of a passenger information system enables passengers to have a picture of the bus system and even the whole traffic system. This reduces uncertainty, and effectively increases passengers' trust in the service.

4.3.3.1.6 High security

The adoption of dedicated bus lanes and the granting of priority at intersections have completely separated the public transport system from other modes of transportation, and reduced possible rear-end and other collisions due to congestion. The establishment of safety systems in vehicles and stations in the fleet management system has further reduced the occurrence of robberies and other violent acts; at the same time, a vehicle-tracking system and traffic accident management system have been adopted. These systems enable timely and prompt rescue when an accident occurs, and increase the personal safety of bus passengers.

4.3.3.1.7   Less pollution and less energy consumption

The adoption of a fleet management system facilitates the best use of existing vehicle resources through effective operation management; the design of the new buses makes it possible to reduce energy consumption and emissions; at the same time, dedicated bus lanes and priority at intersections improve vehicle speed, avoid repeated acceleration, deceleration, and parking in congestion, and can effectively reduce vehicle emissions.

4.3.3.2   Metrobus as a successful practice in Curitiba

Currently, the city of Curitiba has formed a relatively perfect integrated public transport system, which integrates different bus lines physically and operationally into a single network. Physically, the different bus lines are connected in transfer stations, enabling passengers to transfer easily between different lines. Operational integration is based on a single toll system, which allows passengers to transfer freely in all directions, regardless of the length of their journey. Currently, Curitiba's integrated public transport system consists of the following components.

4.3.3.2.1   Line characteristics

The 340 routes and 1550 buses in Curitiba's integrated public transport system cover 1100 km of roads throughout the city. Among these, the dedicated bus lanes are 60 km and the daily mileage of buses is 38,000 km. According to the service characteristics they can provide, these routes can be divided into express lines, branch lines, interregional lines, cycle lines, large station express lines, conventional integrated radiation lines, city central cycle lines, etc.

4.3.3.2.2   Station settings

There are three types of stations in Curitiba's integrated public transport system, namely tubular stations (Fig. 4.5), large bus stations, and traditional stations. The number of tubular stations is 266, and the distance between stations is usually 500−1000 m. Its greatest advantage is that it can greatly speed up passengers boarding and alighting speed. In addition, it can also protect passengers from inclement weather. Meanwhile, the platform-level boarding design and the automatic lifting device at the entrance make it convenient for the elderly and disabled passengers to use the bus system.

Large bus stations are mostly located on the axis of the integrated public transport network, and can be divided into transit function and terminal function. Transit-function stations provide separated boarding and alighting platforms for different lines, and connect these platforms in the form of underground passages, so that passengers can easily transfer. Terminal-function stations are located at the end of the structural axis road, and are

FIGURE 4.5 The inside of a tubular station.

TABLE 4.3 Traffic modes share in Curitiba.

Mode of transportation	BRT + bus	Bicycle	Walking	Car	Motorcycle	Others	Green transport
Share	45%	5%	21%	23%	5%	1%	71%

equipped with large infrastructure to deal with increased traffic between the surrounding areas and the city center.

4.3.3.2.3 Public transport share

Although the vehicles in Curitiba's integrated public transport system run on ordinary roads, they have many characteristics that are similar to a Metro system, such as high-capacity vehicles, ticket sales before boarding, rapid passenger boarding and alighting, and efficient and reliable services, and so they are often referred to as a "road metro system."

Because of the high efficiency and convenience of the public transport system, Curitiba's integrated public transport system occupies a very important position in the urban transport system, with a daily passenger volume of 1.9 million. At the same time, a recent survey shows that 75% of commuters in Curitiba use buses, which is one of the bases for Curitiba reducing the impact of the transportation system on the environment and achieving sustainable development (Table 4.3).

4.3.3.3   Metrobus development experience in Curitiba
In addition to the attractiveness of the Metrobus system, the following factors also play a decisive role in the development of Curitiba's integrated public transport system.

4.3.3.3.1   Integrated transportation planning and urban land use planning
The high-density population needs transportation lines with high transport capacity, and high-capacity transportation lines also need certain traffic demand along the lines to maximize the benefits of the system operation. Based on this idea, Curitiba's public transportation system takes the relationship of land use, road system, and public transportation as the basis of the public transportation system and even the whole integrated transportation system. The interaction and development of these three factors led to the emergence of the main idea of a structural axis, thus providing a flexible, effective, and low-cost traffic solution.

As far as land use is concerned, the formulation of urban planning in Curitiba fully takes into account the principle of matching the intensity of land use with the existing urban structure. Its goal is to adjust the division of residential areas and land use so as to adapt the traffic demand to the social, economic, and urban development. Specifically, the concept of a linear city center is used in the urban planning of Curitiba. The whole city is divided into several districts, each of which determines a special land use management system according to the properties of land use and the intensity of land development.

At the same time, in order to give each district considerable accessibility, the urban road network system is also established hierarchically, which means that the function, characteristics, and capacity of each road in the network are determined to an extent according to its location and importance.

In addition, different land use characteristics also produce different public transport needs. In the vicinity of residential and commercial land that encourages high-intensity land use, the use of dedicated bus roads and dual-articulated buses enables the bus system to achieve a high transport capacity that is consistent with the needs of communities. For residential areas with medium or low population densities, in order to improve operational benefits and the convenience of public transport services, and to enable bus passengers to easily reach other residential areas or traffic nodes, it is necessary to plan routes with relatively low transport capacity and high flexibility.

The land use characteristics of Curitiba linear city make it possible for the BRT system on an axis to operate efficiently. For general city groups, such land use characteristics are not usually present, so it is generally difficult to reach the transport capacity of Curitiba BRT.

4.3.3.3.2 Successful operational management mechanism

The public transport systems of all countries usually run at a financial loss, and it is difficult to ensure a balance of revenue and expenditure. This is not only because a planned public transport network is difficult to better meet the needs of urban public transport, but also because of the poor operation and management of public transport systems. The reason that the Curitiba public transport system has developed so smoothly is also closely related to its successful operation and management mechanism.

4.3.3.3.2.1 The autonomous rights of the Public Transport Administration Curitiba Research and Urban Planning Institute (IPPUC) and Curitiba Urbanization Office (URBS) are at the core of the Curitiba Public Transport Administration. The biggest difference between them and other urban organizations is that they have legal autonomy, they can plan and design urban public transport lines independently, and they have sufficient rights and abilities to implement their decisions, so that their plans can be implemented quickly without prolonged review and discussion, which may led to delays.

4.3.3.3.2.2 Measures of public–private integration Curitiba's public transport system adopts both public and private management measures. The overall urban public transport system is managed by URBS, a public–private joint urban bus company. This company has jurisdiction over 10 private companies. Private companies are licensed from URBS to own public transport vehicles and provide public transport services. They own fleets and are responsible for specific operations. The long-term planning determined by the public management organization avoids the waste of resources caused by excessive attention to local interests, which can make the plan network unreasonable. Private investment in the main construction and operation also can reduce the financial burden on the government to a considerable extent.

4.3.3.3.2.3 Separation of operation and ticket system The integrated public transport system in Curitiba is managed by URBS to accomplish a specific operation. The state government provides many conveniences for private companies, such as providing guarantees for loans from banks to private companies. The ticketing system is run by an integrated bus system foundation. This foundation has a special organization to study and develop the ticket system; the management system consists of the municipal government to control the operation mileage, private companies to accomplish the operation mileage, and the foundation to sell tickets.

4.3.3.3.3 Flexible construction measures

Since Curitiba's public transport system uses public transport vehicles that can travel on ordinary roads, it is possible to build and implement public transport networks in batches and phases. In the case of technical and financial support, different parts of the system can also be upgraded according to demand. The implementation process of batch and phased construction is also convenient for identifying problems in the planning of the public transport network through continuous monitoring during use, so that the original scheme can be amended continuously in the future planning and construction process, forming a perfect feedback process, avoiding the huge losses caused by the one-time decision-making mistakes of large public facilities.

4.3.3.3.4 Adoption of relevant transportation management measures

4.3.3.3.4.1 Priorities of bicycles and walking Bicycles and pedestrian areas are an integral part of the road network and public transport system. Curitiba has vigorously built bicycle lanes, while the downtown business district also has large pedestrian areas.

4.3.3.3.4.2 Parking measures Parking in Curitiba is restricted by clear control measures and is strictly monitored and enforced. For example, the municipal government of Curitiba adopts reasonable parking strategies to reduce roadside parking, it stipulates that taxis must stop at taxi parking stations, and signs prohibiting parking are set at intersections to ensure the smooth passage of public transport at roads and intersections.

4.3.3.3.4.3 Economic policies to encourage the use of public transport To encourage citizens to use the public transport system to travel, Curitiba municipal government stipulates that: the elderly over 65 years old and the children under 5 years old can take public transport without buying tickets; for Curitiba citizens with wage income, if the cost of public transport exceeds the wage, the excess part shall be subsidized by the government; for the poor living in the poor areas, they can use the garbage cleaning in exchange for bus tickets.

Further reading

<http://ditu.google.cn/maps?client = aff>.
<http://nemopa.blog.sohu.com/148901704.html>.
<http://hi.baidu.com/starbird/item/26e7aaf63eb7fdca531c269d>.

Chapter 5

Tokyo, Japan

Chapter Outline

5.1 An overview of Tokyo

Tokyo, is the capital, and also the political, economic, and cultural center of Japan. Tokyo is also one of the world's most famous megacities. It is composed of 23 districts, 26 cities, 5 towns, and 8 villages, with a population of 13.28 million (2013) in the Tokyo area of 2187.58 km^2. The core area of

Eco-Cities and Green Transport. DOI: https://doi.org/10.1016/B978-0-12-821516-6.00005-9

Tokyo, which consists of 23 districts, is called the "Tokyo Metropolitan Area" with an area of 622.99 km², a population of 9.055 million people (August 2013), and a population density of 14,563 inhabitants per square kilometer. Chuo city, Chiyoda city, and Minato city are the three core districts of Tokyo and are known as the "three districts of the capital center," which is the earliest urban center of Tokyo.[1] The population of the Tokyo Metropolitan Area varies greatly by day and night, especially in the high-density commercial centers such as Shinjuku, Ginza, and Tokyo station. In this area, the daytime population can reach more than nine times the night population. Therefore the Tokyo Metropolitan Area is densely populated and lively during the day, but after midnight it becomes almost an empty city. This is mainly due to the high land and housing prices in the central city of Tokyo, resulting in the area being used primarily for the development and utilization of public buildings such as finance, commerce, and enterprises. Most residents reside in the suburbs, exurbs, and even the surrounding counties and cities; they work in the center of the Tokyo Metropolitan Area during the day but return to their suburban homes in the evening, forming a strong tidal commuter flow.

Tokyo has extremely strong radiative effects on the economy, industry, population distribution, and mobility of surrounding cities and counties. Especially in the morning and evening tidal peaks, the traffic flow shows obvious characteristics of the metropolitan area. Therefore when it comes to Tokyo, we must first clearly distinguish the four concepts of the capital circle, Tokyo capital circle, Tokyo, and the Tokyo Metropolitan Area. According to the statistics of the capital circle and the statistics of the White Paper on the Capital Circle issued by the Ministry of Land, Infrastructure, Transport and Tourism of Japan, the capital circle includes "one city and seven counties," namely Tokyo, Ibaraki, Tochigi, Gunma, Saitama, Chiba, Kanagawa, and Yamanashi. The Tokyo Metropolitan Area covers only the "one city and three counties" in Tokyo and the capital city, which are the most developed and geographically close to Tokyo, including Tokyo, Saitama, Chiba, and Kanagawa. The two major cities of Kawasaki city and Yokohama city are close to Kanagawa Prefecture. As of April 2012, the population of the Tokyo Metropolitan Area was 37.126 million [1]. The area and population data of Tokyo, Tokyo Metropolitan Area, and Tokyo capital circle are shown in Table 5.1.

Tokyo capital circle is the world's most populous and economically largest metropolitan area, surpassing the world's second largest—New York economic circle. For example, in 2008 Tokyo's gross national product reached 89.7 trillion yen, or about $1 trillion, 60% higher than New York's $610 billion in gross national product [2]. In 2010, the nominal GDP of the Tokyo

---

1. These data come from Wikipedia and organize the statistics of the Japanese government into the "Capital Circle," "Tokyo," and "Tokyo Metropolitan Area."

TABLE 5.1 The populations of Tokyo, Tokyo Metropolitan Area, and Tokyo capital circle.

	Tokyo capital circle (one city and three counties)	Tokyo	Tokyo Metropolitan Area (23 districts)
Land area (km^2)	13,559.21	2188.67	622.99
Population (10,000 people)	3712.6 (2012 data)	1328.2 (2013 data)	905.5 (2013 data)
Population density (persons/km^2)	2738	6070	14,563

Source: Data from Tokyo Metropolitan Government Statistics and Wikipedia.

capital circle area was 160 trillion yen [3], which was about 1.6 trillion US dollars. In the same year, the gross national product of the New York metropolitan area was 1.36 trillion US dollars [4].

5.2   Urban structure and land use

Until the end of the 1950s, Tokyo was a single-center city. Many administrative agencies, headquarters of large enterprises, financial organizations, and commercial organizations gathered in the three core districts of Chuo city, Chiyoda city, and Minato city, which formed the earliest urban center in Tokyo. After World War II, with the recovery and rapid growth of the Japanese economy, the commercial functions of the three core districts of Tokyo were further developed, and the urban functions were sharply concentrated here, causing soaring land prices and housing shortages. The core residents moved to the suburbs, and there was a sharp increase in commuter traffic. This caused an increase a series of urban issues. In order to restrain unrestricted growth at the center and to solve the urban problems caused by it, in 1958, the Tokyo Metropolitan Government began to establish the three subcenters of Ikebukuro, Shibuya, and Shinjuku to guide the transition of the city from a single center to a multicenter structure. In 1982, construction of the subcenters of Ueno-Asakusa, Osaki, and other cities began. By the end of the 20th century, Tokyo had built seven subcenters along the circular track Yamanote Line. The Tokyo Metropolitan Area formed a multicenter urban structure, as shown in Fig. 5.1. However, the population and industry were concentrated rapidly due to the rapid growth of Japan's economy in the 1970s and 1980s. The deputy center and the three core districts are concentrated along a line and the inner side of the Yamanote Line, with a diameter

FIGURE 5.1 Tokyo Metropolitan Area multicenter urban structure. *Created according to the CraftMap basemap, http://www.craftmap.box-i.net/sozai.php?no = 1009_5.*

of about 10 km, and only 5 km distance between each other. Nowadays, the entire Yamanote Line has been integrated into a high-density and large-scale urban center. A large number of jobs have been concentrated here. At the same time, due to the rising land price in the central city, residents began to migrate to the suburbs, creating a mode of living in the suburbs and working in the city center. To solve the problem of housing for a large number of residents, the government and enterprises have built many new cities around Tokyo. Residential houses have also been built in the surrounding cities and counties of Tokyo, forcing the urban green belt to retreat. In the 1980s, the green belt of Tokyo retreated 50 km away from the central city [5] and has not avoided disorderly expansion.

Rapid economic growth has raised the initially low price of land in urban centers, and the land use has been very significant. The high-density development of public buildings such as commercial, financial, and administrative offices in urban centers, and the migration of residential buildings to the suburbs, have forced Tokyo to build a strong radial railway to connect residences and urban centers. This series of joint development and construction activities has made Tokyo Metropolitan Area into an obvious "circle structure." From the center to the periphery, the division of functions in the urban area is clear. The separation of occupation and residence has serious implications. The city center is high-rise and there is a lack of green plant areas. Some people use the term "Tokyo desert" to describe this high-intensity development economic city. While some use the phrase "commuting hell" to describe the daily commutes of Tokyo Metropolitan Area residents. Statistics show that the average one-way commute time for people living in the metropolitan area using railway in 2010 was 68 minutes [6]. The one-way commute distance can be up to 60 km, and the one-way commute time up to 3 hours.

5.3   Motorization and traffic demand characteristics

In 2008, Tokyo conducted a survey of residents' travel needs in the metropolitan area of Tokyo by selecting Tokyo and four prefectures in the capital circle (Tokyo, Kanagawa, Saitama, Chiba, and the south of Ibaraki) as survey areas. At the time of the survey, the Tokyo Metropolitan Area had a population of 36.02 million. The average one-way trip time was about 34 minutes. The average daily travel volume was 84.89 million (2.4 times greater than Paris), and the average daily travel frequency was 2.37.

Fig. 5.2 shows the sharing rate of residents' travel modes in the Tokyo capital circle, Tokyo Metropolitan Area, and some cities. As can be seen from the data in this figure, the total railway utilization rate in the Tokyo capital circle is 30%, the bus sharing rate is 3%, and motor vehicle travel accounts for 29%. The full-time residential travel ratio of railways to buses is 51%, and motor vehicle travel only 11% in Tokyo Metropolitan Area.

This shows that in the Tokyo Metropolitan Area, more travelers use public transportation, and the utilization rate of personal motor vehicles is far lower than the overall level for the Tokyo Metropolitan Area. The railway sharing rate of commuter traffic in the Tokyo Metropolitan Area is as high as 79%. The daily traffic volume of transportation hubs such as Tokyo

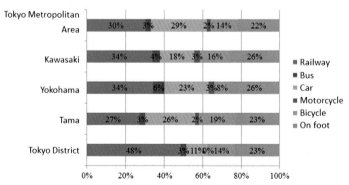

FIGURE 5.2   Tokyo capital circle, Tokyo Metropolitan Area, and some cities total travel mode sharing rates.

TABLE 5.2 Traffic mode sharing rates of the Tokyo Metropolitan Area.

Transportation mode	Walking	Bicycle	Rail transit	Bus	Private car	Motorcycle	Total green transport
Traffic sharing rate (%)	23	14	48	3	11	1	88

station and Shibuya station has reached millions of passengers. The daily passenger traffic of Shinjuku transportation hub is more than 3 million passengers, ranking first in the world. This shows that the proportion of public transportation use in the Tokyo Metropolitan Area is very high, as shown in Table 5.2, which is a typical city dominated by public transportation [7].

The Tokyo Metropolitan Area government is the center of business, finance, entertainment, and culture in the Tokyo Metropolitan Area. The office and commercial facilities are highly concentrated. Most citizens live on the edge of the district or other cities and counties outside Tokyo, forming a clear function for the city. Most citizens live on the edge of the district or other cities and counties outside Tokyo, forming a clear function of the city. There is a clear pattern of separation. To cope with the strong commuting traffic demand, the Tokyo Metropolitan Area and Tokyo capital circle have an impressive railway transportation system that has been in existence for a long time, making the rail transit system of the Tokyo Metropolitan Area highly developed. However, it has not been able to solve the severe commuting traffic problem. As the center of the metropolitan area, the Tokyo Metropolitan Area has attracted the most travel. Of these, the Tokyo Metropolitan Area is most closely connected with the surrounding Tama, Saitama, the northwest of Chiba, Kawasaki, and Yokohama in Kanagawa. There is a large amount of cross-regional commuting. In 2010, the total number of commuters in Tokyo was about 9.43 million, of which 3.1 million entered from peripheral counties, and 3.37 million entered from outside the district. The peripheral counties are dominated by Saitama, Kanagawa, and Chiba Prefectures in Tokyo. The one-way average travel time by rail is more than 1 hour [8]. This shows that the distribution of occupations and residences in Tokyo is seriously unbalanced, the urban structure is unsuitable, and commuting traffic is time consuming.

In 2011, the total number of motor vehicles in Tokyo was 3.23 million, with about 245 motor vehicles for every 1000 people. Of these, the number of motor vehicles in the Tokyo Metropolitan Area was 2 million and the number of motor vehicles per 1000 population was about 220. The average number of motor vehicles owned by 1000 people in other parts of Japan and other OECD countries is more than 400. The developed public transportation system in Tokyo has helped control the total number of motor vehicles.

5.4   Rail transit development

Traffic organization in the Tokyo Metropolitan Area relies heavily on rail transit systems throughout the city and suburbs. The Tokyo Metropolitan Area is densely populated with a number of operating companies, primarily including ordinary railways operated by Japan Railway Company and private railways operated by a number of private companies. The subway is operated by two companies, Tokyo Metro and Toei Transportation. In 2011, the total length of the rail network in Tokyo was 1089.1 km. The Japan Railways system had an extent of 391 km [9] and the average daily passenger traffic was about 8.84 million passengers. The private railway operating extent was 406.8 km and the average daily passenger traffic was 7.99 million. The subway operating network was a total of 291.3 km and the average daily passenger traffic about 8.5 million passengers.[2] The operating mileage and average daily passenger traffic of each rail system and their proportions are shown in Fig. 5.3.

The JR Line has a wide range of services. The main rail transit lines in the outer suburbs of Tokyo are mainly operated by the JR Line, with an average station spacing of 2.77 km. The private railway mainly serves the Tokyo metropolitan area and suburbs. The average station spacing is 1.09 km, which is less than the average station spacing of the JR Line, and the line load intensity is also high. The subway lines of the Tokyo Metro and Toei are mainly operated in the Tokyo Metropolitan Area, with short site spacing and high line load intensity. The operation of the JR Line in Tokyo in 2011 is shown in Table 5.3. The basic conditions of the operation of the private railway lines are shown in Table 5.4. The operation of each subway line operated by the Tokyo Metro Corporation is shown in Table 5.5. The operation of each subway line operated by the company is shown in Table 5.6.

5.5   Shinjuku transportation hub and surrounding land integration development case study

5.5.1   Development history of Shinjuku transportation hub

The Shinjuku transportation hub is located in the western part of Tokyo's central city. It is about 10 km from the earliest urban core areas such as Tokyo station and Ginza. It is the gateway to the Tokyo Metropolitan Area and the western suburbs, cities, and counties, and the western counties of the Tokyo capital circle. From the earliest underdeveloped suburban railway station to becoming today's modern integrated transportation hub with the world's highest passenger traffic, Shinjuku Station has gone through more than 100 years of development history.

---

2. Based on statistics from Tokyo Metro and Toei Subway. http://www.tokyometro.jp/corporate/enterprise/passenger_rail/transportation/lines/index.html;   http://www.kotsu.metro.tokyo.jp/information/service/subway.html.

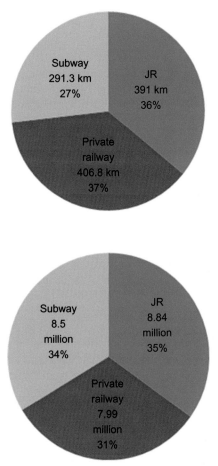

FIGURE 5.3   Distribution of operating mileage and average daily passenger traffic of various rail transit systems in Tokyo.

TABLE 5.3 Operation of the JR Line in Tokyo in 2011.

Total number of railway lines (bar)	Line length (km)	Number of stations	Average station distance (km)	Average daily passenger traffic (10,000 people)	Line load intensity (10,000 passengers/ km day)
14	391	141	2.77	883.8	2.26

Source: Data from The Statistical Yearbook of the Tokyo Capital in Pingcheng 23 Years.

TABLE 5.4 Basic conditions of private operating companies in Tokyo in 2011.

Operating company	Total number of railway lines (bar)	Line length (km)	Number of stations	Average station distance (km)	Average daily passenger traffic (10,000 people)	Line load intensity (10,000 passengers/km day)
Odakyu	2	26.5	23	1.15	88.66	3.35
Seibu	10	93.4	71	1.32	138.52	1.48
Tokyu	7	60.0	75	0.80	179.09	2.98
Keihin Kyukou	2	18.3	20	0.92	41.29	2.26
Keio	2	77.6	66	1.18	161.80	2.09
Keisei	3	23.9	19	1.26	35.11	1.47
Tobu	4	28.2	28	1.01	84.45	2.99
Hokuso	1	2.0	2	1.00	3.40	1.70
Takao Tozan	2	1.9	4	0.48	0.53	0.28
Tokyo Monorail	1	17.8	11	1.62	12.10	0.68
Mitake Tozan	2	1.1	4	0.28	0.15	0.14
Yurikamome	1	14.7	16	0.92	8.32	0.57
Tokyo Waterfront Area Rapid Transit (TWR)	1	12.2	8	1.53	0.00	1.62
Tamatoshi Monorail	1	16.0	19	0.84	12.30	0.77
Metropolitan Intercity Railway	1	13.2	7	1.89	13.75	1.04
Total	40	406.8	373	1.09	799.23	1.96

Source: Data from The Statistical Yearbook of the Tokyo Capital in Pingcheng 23 Years.

TABLE 5.5 Operation of metro lines of the Tokyo Metro Company in 2011.

Railway route name	Line length (km)	Number of stations	Average station distance (km)	Average daily passenger traffic (10,000 people)	Line load intensity (10,000 passengers/ km day)
Ginza	14.3	19	0.75	71.17	4.98
Marunouchi	27.4	28	0.98	84.18	3.07
Hibiya	20.3	21	0.97	84.85	4.18
Tozai	30.8	23	1.34	76.45	2.48
Chiyoda	24.0	20	1.20	89.66	3.74
Yurakucho	28.3	24	1.18	66.96	2.37
Hanzomon	16.8	14	1.20	61.75	3.68
Namboku	21.3	19	1.12	34.12	1.60
Fukutoshin	11.9	11	1.08	21.43	1.80
Total	195.1	179	1.09	622	3.19

Source: Data from Tokyo Metro Co., Ltd.

TABLE 5.6 Operation of metro lines of the Tokyo Metropolitan Subway Company in 2011.

Railway route name	Line length (km)	Number of stations	Average station distance (km)	Average daily passenger traffic (10,000 people)	Line load intensity (10,000 passengers/ km day)
Asakusa	18.3	20	0.92	60.77	3.32
Mita	26.5	27	0.98	55.51	2.09
Shinjuku	20.7	20	1.04	62.21	3.01
Oedo	40.7	38	1.07	78.29	1.92
Total	106.2	105	1.01	225.19	2.12

Source: Data from The Statistical Yearbook of the Tokyo Capital in Pingcheng 23 Years: Subway transportation results.

Shinjuku station was first used on March 1, 1885. The first north—south railway line in Tokyo, the Shinagawa line (Akasu station to Shinagawa station) was opened. Shinjuku is a stop on this line. Shinagawa Line is also the predecessor to some sections of the JR Yamanote Line today. On April 11, 1889, the Kaiwu Railway (the predecessor of the section of the JR Central Line) was opened from Shinjuku to the Tachikawa Line. On May 1, 1915, the Keio Electric Track (now Keio Electric Railway) Shinjuku station was opened. On April 1, 1927, the Odawara Express Railway (now Odakyu Electric Railway) Shinjuku station was opened, enabling Shinjuku to gradually develop into a prosperous neighborhood. In 1959, the Shinjuku subway station was opened. The early Shinjuku station was located in the suburbs. When it first opened, the passenger flow was very small, the land development intensity around the site was very low, and the development speed was slow. Until the 1930s, with the gradual expansion of the size of Tokyo and the gradual introduction of Shinjuku by private railway lines, the traffic volume at Shinjuku station began to show a continuous growth trend, and the surrounding land development intensity gradually increased. After World War II, Japan's economy developed rapidly. In 1958, the Tokyo Metropolitan Government established Shinjuku as a subcenter of urban development, planned the west side of Shinjuku station as a super-tall building area, and carried out high-intensity development around the site, focusing on the construction of a number of administrative public facilities including business, culture, and entertainment ventures, causing the number of passengers in Shinjuku station to increase rapidly.

5.5.2   Overview of Shinjuku transportation hub

At present, the Tokyo Metropolitan Line, the Shonan Shinjuku Line, the Narita Express Line, the Central Line, the Yamanote Line, and the Sobu Line (the main lines are directly connected to the Central Line) operated by JR East Japan; the Keio Line and the Keio New Line operated by Keio Electric Railway; Odakyu Line operated by Odakyu Electric Railway; Marunouchi Line operated by Tokyo Metro; 12 tracks of Shinjuku Line and Oedo Line operated by Toei Subway enter Shinjuku transportation hub, including JR East Japan (except JR Yamanote Line) and Keio Electric Railway. Most of the Odakyu electric railway leads to suburbs, cities, and counties outside Tokyo, and it is responsible for passenger traffic between the Tokyo suburbs and the Tokyo Metropolitan Area. The subway line mainly carries passenger traffic in the Tokyo Metropolitan Area. The above five rail transit operators set up separate Shinjuku stations, however, these stations are connected to each other. Passengers can transfer directly between different lines and different stations, and it purchasing tickets is very convenient. Even the track lines of different operating companies are directly connected, for example, between the Keio Electric Railway New Line and the Toei Subway Shinjuku Line, passengers at Shinjuku station do not need to transfer, as there is a direct connection.

The various orbital routes originating or passing through Shinjuku station connect the Tokyo Metropolitan Area with many "sleeping cities" in the surrounding areas of the city, creating 3.46 million daily average passengers, which is a world record for an integrated transportation hub. The daily average number of passengers on each track line of the Shinjuku transportation hub in 2010 is shown in Table 5.7.

According to the daily average passenger traffic of each rail transit line shown in Table 5.7, the proportion of each rail line to the traffic at Shinjuku station can be calculated, as shown in Fig. 5.4. It can see that the passenger traffic of JR East Japan Railway, Keio Electric Railway, Odakyu Electric Railway, and nine other lines accounts for 82% of the total passenger traffic of Shinjuku station, and most of these lines (except the JR Yamanote Line) take traffic between Tokyo and other counties and cities, showing that the bulk of the passenger flow of Shinjuku station is for external traffic passengers. It can also be seen that for Tokyo, Shinjuku station plays an important role as a transit hub for internal and external traffic.

5.5.3 Set up of three-dimensional channels and clear signs in the transportation hub

Every day, more than 3 million passengers pass through the Shinjuku transportation hub. The passenger flow is high and continuous, but it happens smoothly and efficiently. This is mainly due to the careful design of the perimeter of the station, the three-dimensional walking paths raised off the ground, and clear signage. Within an area of about $2\,km^2$, more than 100 exits have been set up in the Shinjuku integrated transportation hub, so that travelers are able to use the rail transit for all destinations without leaving the hub. This design principle of tightly combining rail transit entrances and exits with surrounding areas means that the rail transit is well served, and has achieved great success in facilitating travelers and promoting the use of green transportation.

Fig. 5.5 shows a number of indicating signs above the central passage in the hub. Clear information is given by a combination of different colors, numbers, characters, and symbols. Passengers are able to see the tracks to the stations at any time and from any location, including the direction of travel. Fig. 5.6 shows the names and exit locations of the buildings around the hub. The passage design also takes into account the accessibility of surrounding buildings from the hub. A number of exits can be used. From the hub, passengers can reach the destination building directly along the underground passage. The large-scale electronic display screen promptly shows the boarding time and platform for each line, train to different directions, as shown in Fig. 5.7. The advanced hardware facilities, high-level management methods, and traffic organization design of

TABLE 5.7 Daily average passenger traffic distribution of each track line in Shinjuku Station in 2010.

Railway route name	JR East Japan	Tokyo Metro	Toei transportation Shinjuku line	Toei transportation Oedo line	Odakyu electric railway	Keio electric railway	Total
Average daily passenger traffic (person/day)	1,473,430	212,426	266,208	124,166	483,304	715,950	3,275,484

Source: Data from These data are based on the number of passengers at each station published by Wikipedia and various railway companies.

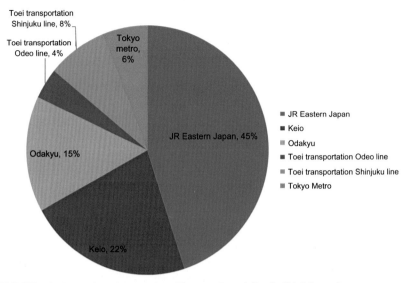

FIGURE 5.4 Proportion of passenger traffic on each track line in Shinjuku station.

FIGURE 5.5 Directional signs in the central passage of the Shinjuku transportation hub.

the Shinjuku transportation hub make this, the world's largest transportation hub with the largest passenger flow and the most track lines, safe, orderly, and efficient.

FIGURE 5.6   Directional signs to the surrounding buildings in the Shinjuku transportation hub.

FIGURE 5.7   Electronic travel information in the Shinjuku transportation hub.

5.5.4   Integrated development of rail transit stations and commercial facilities

Five rail transit companies—JR East Japan Railway, Keio Electric Railway, Odakyu Electric Railway, Tokyo Metro, and Toei Subway—have set up sites in Shinjuku, which are connected to each other to form the Shinjuku transportation hub. These sites are not purely transportation facilities, but also offer a comprehensive building integrating transportation hub, commercial shopping, and offices. The station, which is usually used as a transportation hub function, is located underground. The building area above the station is used as a shopping center, retail, catering, and office building for general development and management. It provides convenient conditions for travelers to enjoy meals, shopping, etc., making it a comprehensive transportation hub.

Many rail lines and integrated transportation hubs create strong demand. Shinjuku has become a gathering area for a number of people. People travel there to work, but then create a series of needs for transportation, catering, shopping, entertainment, etc. In addition to its comprehensive development and utilization, the land around the Shinjuku transportation hub has also been developed for larger scale commercial and office facilities, and these buildings and the hub building are organically connected through underground passages, as shown in Fig. 5.8. Clear signage is used throughout the station. The passages are clean and bright, and unaffected by natural sunlight, rain, and snow, creating a warm and pleasant walking environment. After alighting at Shinjuku station, travelers can easily access the business districts and office buildings through underground passages that are well signed.

FIGURE 5.8 Underpass connecting Shinjuku station with the surrounding commercial and office facilities.

5.5.5 High-intensity development of Shinjuku as the deputy center of Tokyo—coexistence of advantages and disadvantages

Since the 1960s, as the deputy center of Tokyo, Shinjuku has carried out high-intensity development of the surrounding areas of the transportation hub, driven by the huge rail transit hub. As shown in Fig. 5.9, the four main blocks of Shinjuku, Nishi Shinjuku, Kabukicho, and Yoyogi are within 500 m of the center of Shinjuku station and have become high-density business districts and administrative offices for Tokyo. Economic development has brought vitality and provided a good foundation for development. Nishi Shinjuku has become a key area for the development of the deputy center, forming a super high-rise building area represented by the Plaza Hotel, the Shinjuku Sumitomo Building, and the Tokyo Metropolitan Government Building, as shown in Figs. 5.10 and 5.11. These super high-rise buildings are a symbol of the rapid development of the Japanese economy and the rapid improvement in the level of modernization. Thanks to the numerous railway transportation lines, the integrated hub has convenient connections and evacuation functions, helping these large administrative and commercial facilities to function efficiently. People working here, arriving from the station, can reach the offices or shopping center smoothly within 5 minutes by accessing the underground passage.

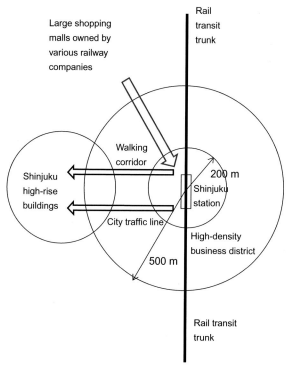

FIGURE 5.9 Schematic diagram of land use around Shinjuku station.

FIGURE 5.10 Aerial view of the Shinjuku high-rise buildings from the Tokyo Metropolitan Administration Building.

FIGURE 5.11 High-rise buildings that can be reached by walking from Shinjuku station via the underground passage in about 5 min.

This integrated development mode of the transportation hub and surrounding land reduces people's travel time, and enables rapid rail transit, and convenient integrated hub function, prompting people to abandon the use of motor vehicles and make extensive use of rail transit, which not only meets people's travel needs, but is also in line with the development of green transportation. The successful model of integrated development of Tokyo rail transit and its surrounding land is an important example to others.

At the same time, the high-intensity development of Shinjuku's deputy center has also brought about a prominent problem of inadequate supply of land, meaning there are almost no residential houses here. In the area around Shinjuku station (with a radius of about 1 km and an area of about 3 km^2), the proportion of day to night population can be as high as 9.04:1 [10]. In the three blocks of Shinjuku-Sanchome, Nishi-Shinjuku 1-chome, and Nishi-Shinjuku 2-chome, which are the closest to Shinjuku station, the area is only 0.84 km^2, and the proportion of day to night population is as high as 294.45:1 [10]. It can be seen that the administrative functions and business of Shinjuku's deputy center function and living function are highly unbalanced. During the day, there are crowds of workers, and there is a clear tidal traffic flow in the morning and evening, leaving empty and quiet high-rise buildings at night.

The goal of being an eco-city with green transportation is to meet the needs of people's daily work, life, and travel, and to provide comfortable and fast transportation services while reducing the negative impacts on the environment. From the perspective of traffic demand characteristics, Tokyo's urban structure and land form are essentially megacities with a single center and highly concentrated central functions. In response to these urban structural and traffic demand characteristics, Tokyo has a well-developed rail transit system and an integrated transportation hub, which greatly facilitate people's work and life travel, achieving high-efficiency urban traffic in modern megacities. However, the urban structure of a single center and the excessive concentration of urban functions have also brought about drawbacks to megacities that are difficult to solve, such as long-distance commuting. The resulting increase in urban operating costs and the waste of commuters' travel time and energy are also enormous. The lessons from Tokyo in urban development are worth learning from.

5.6 Public transportation guides urban development: a case study of a new city construction around Tokyo

5.6.1 The transit-oriented development mode and new city construction in the Tokyo Metropolitan Area

The transit-oriented development (TOD) model is a "public transportation-oriented development" model. The concept of TOD was first proposed in

1992 by Peter Calthorpe, a representative of American neo-urbanism. After World War II, American society experienced a series of problems such as the car being the main means of transportation, resulting in unrestricted spread to the suburbs, urban population migration to the suburbs, reduced land use density, a sharp increase in energy consumption, and an impact on the ecological environment. After nearly half a century of development, people have finally realized the unsustainability of this development model, and began to seek ways to reduce energy consumption, protect the ecological environment, and build a warm and harmonious ecological city and community. American scholars first proposed the TOD model. The core concept is to use public transportation such as rail transit and bus trunks as the main modes of travel, with a bus stop as the center and a radius of 400−800 m to build a city integrating work, business, culture, education, and residence. Travel within the community is mainly based on walking. Travel between the community and the outside is mainly based on public transportation such as rail or bus, and the organic coordination mode of compact development of each city group is realized.

Although the TOD model was only proposed in the 1990s, Japan began to use this model for land development and construction in the 1960s. Due to the extremely limited land resources in Japan, the postwar economy developed rapidly, and a huge number of people gathered in large cities with developed industrial and commercial opportunities. In particular, the population density of the three metropolitan areas of Tokyo, Osaka, and Nagoya increased dramatically. To address the needs of the urban population for living and commuting, from the 1950s and 1960s Japan's radial railways were built from the center of the city to the periphery. Many new cities were built around railway stations. People lived in the new city and took rail traffic to the city center to work. This land development and utilization model has strong consistency with the TOD model later proposed in the United States.

There are many cases in the metropolitan area of Tokyo that guide the construction of new cities by public transportation. Tama New Town, Hachioji, Tachikawa, and Shinsan township are representative examples of new city development around rail transit sites.

5.6.2 Tama New Town

Tama New Town is located in the Tama hills to the west of the Tokyo Metropolitan Area, 25−35 km from Tokyo station. In the 1960s with the rapid growth of the Japanese economy, jobs in the Tokyo Metropolitan Area were highly concentrated and there was a serious shortage of housing. At the same time, due to the high land price in the Tokyo Metropolitan Area, it was difficult for residential construction agencies to obtain land for residential construction within the district. At that time, the Tama area was in a state of disorderly development of suburban golf courses and small residential areas.

To solve the housing problem of those in the Tokyo Metropolitan Area, the government began to consider the development and construction of large-scale residential areas with good living conditions. In the early 1960s the Tokyo Metropolitan Government first proposed the residential development and construction plan for the South Tama area (1960−61), and planned to build a collective residential area to house about 150,000 people in about 16 km^2, with Tama village and Daocheng village at the center. After several years of discussion, a revision of the framework of the new city construction, and the approval of the local town government involved in the new city, the Tama New Town construction project was approved by the Ministry of Construction of Japan in December 1965.

Tama New Town is a residential development center with a planned area of about 30 km^2 and a planned population of 342,200. The planned area is about 14 km long from east to west and about 3 km wide from north to south. Taking the rail transit site as the core, the development of residential and park green space is the main focus, forming a new city based on a rural/urban ideal. Table 5.8 is the land use plan for Tama New Town in 1966. It can be seen from the planning data in this table that the sum of residential and park green space amounts to about 55%, while commercial land is small, and commercial and business-specific facilities use only about 5%. There are few jobs in the new city. Employed people commute to work in the Tokyo Metropolitan Area. The new city is mainly a residential area and has become a so-called "sleeping city." In 2010, the population of Tama New Town reached 216,000, and there remains a trend toward continuous growth.

Considering that the development of Tama New Town would bring a number of residents to work in the Tokyo Metropolitan Area, generating strong commuter traffic demand, the planning and construction of the rail

TABLE 5.8 Land use planning of Tama New Town in 1966 [11].

Land type	Planned area (ha)	Proportion (%)
Residential	785.6	35.3
Park green space	432.9	19.4
Educational	212.6	9.6
Others public facilities	4.8	0.2
Roads	421.7	19
Commercial	77.6	3.5
Specific business facilities	61.2	2
Other public welfare	229.2	10.3
Total	2225.60	100

transit system was carried out at the start of planning the Tama New Town. In 1964, the "Tama New Town Transportation Plan" prepared by the "Southern Tama Regional Transportation Plan Investigation Committee," which is based in the University of Tokyo, discussed the importance of building a rail transit system between Tokyo and Tama New Town. The method put forward a suggestion that the extension of the original line construction of the two track operation companies Odakyu Electric Railway and Keio Electric Railway would be extended to the new city to achieve supporting rail transit for the new city both efficiently and economically. In the following 10 years, this track construction was synchronized with the development of the new city. The two track lines have eight stations in the Tama New Town area. Tama Center station is the most important site in the new city. Odakyu Tama Line opened in June 1974 reaching as far as Yongsan station. In 1975, it was extended to Tama Center station. The Jingwang Sagami Line was opened as far as the Tama Center station in October 1974. At the same time as the track construction, the residential construction department developed residential areas, park green spaces, schools, and commercial facilities around the site.

Fig. 5.12 shows the Tama Center station surrounding area in 2005. Nearly 30 years since its development and construction, the city around the

FIGURE 5.12 The appearance of Tama Central station in 2005. ① Keio Line, Odakyu Line Tama Center station; ② Tama area Monorail Tama Center station; ③ Domo Warner Movie city; ④ Shin Kong Mitsukoshi department store. *From http://8843th.at.webry.info/200611/article_2.html.*

station is now bustling with commercial facilities, with the Odakyu Electric Railway, Keio Electric Railway, and Light Rail all present. The station has been set up, and there are entertainment facilities and shopping centers around the station, which have created an integrated transportation hub for Tama New City.

At present, there are three lines, Jingwang Sagami Line, Odakyu Tama Line, and Tama Monorail in the Tama New Town area, which are responsible for commuter passenger transportation between the new city and the Tokyo Metropolitan Area. It takes about 40 minutes to travel from Shinjuku's deputy center by rail transit to Tama Center station, which is shorter than the average commuter time in the Tokyo Metropolitan Area.

5.6.3   Tachikawa city and Hachioji city

Tachikawa city and Hachioji city are located to the west of the Tokyo Metropolitan Area and are on the north side of Tama New Town. The relative position of the three cities is triangular, and they belong to the Tama area of Tokyo. Tachikawa is 28 km from Shinjuku, while Hachioji is 40 km from Shinjuku. Since the two cities are located on the JR Main Line, the main trunk railway from Tokyo to the central part of Japan, there is a strong rail transit between the two cities and the Tokyo Metropolitan Area. It has greater vitality in terms of industrial development and employment.

Tachikawa station is the largest rail transit hub in the Tama area. As early as 1889, with the opening of the predecessor to the JR Chuo Main Line, the Asakawa Railway Line was opened. At present, the three rail transit lines of JR Chuo Main Line, Nanbu Line, and Ome Line pass through Tachikawa station, and the transfer to the Tama Area Monorail Line is completed through Tachikawa North station and Tachikawa South station. The surrounding areas of Tachikawa station have been relatively well developed. Most of the area within about 500 m of the rail transit station is multistory, high-rise buildings. There are some commercial and office facilities, forming a business district centered on the station. The expansion includes residential areas and park green spaces, with the houses being mainly low-rise buildings. The land use development level around Tachikawa station is distinct, the functional division is prominent, and the occupational residence is well balanced. As shown in Fig. 5.13, to the north and south of Tachikawa station are Tachikawa South station and Tachikawa North station, respectively, with Tachikawa station being an integrated rail transit hub. Fig. 5.14 shows the development status of commercial facilities around Tachikawa station.

Hachioji city, located to the west of Tachikawa city, is also located on the JR Chuo Main Line. It is farther from Tokyo than Tachikawa city, however, due to the strong JR railroad link connecting with Tokyo, travel between Tokyo and Shinjuku is very convenient.

FIGURE 5.13 Land use pattern around Tachikawa station. ① Tachikawa station; ② Tachikawa South station; ③ Tachikawa Kita station. *From http://jpp-k.com/2011_09_10_1540.html aerial photography photo modification.*

FIGURE 5.14 Development around Tachikawa station. *From http://www.city.tachikawa.lg.jp/ cms-sypher/open_imgs/service/0000000116_0000002146.JPG.*

Because Hachioji city is far from the city center of Tokyo, more land available for development, and it is close to the mountainous areas of central Japan. The environment is elegant and good natural ecology. Rail transit enables easy travel to and from the Tokyo Metropolitan Area, thus attracting many commuters. Educational and scientific research institutions have settled here, and there are many primary and secondary schools within about 2 km from Hachioji station, and about 5 km from the station a university has been built. Fig. 5.15 shows the Soka University campus about 5 km from Hachioji station.

Hachioji station is also a historic rail transit site that was first put into use in 1889. The station building itself is integrated with department stores to form a comprehensive building integrating rail transit sites with commercial facilities, as shown in Fig. 5.16. A relatively mature business district has also formed around the station. Fig. 5.17 shows the shopping street near Hachioji station. There are mature residential communities and research and education industrial zones within 500 m of the station. About 100,000 commuters travel to and from Tokyo daily and about 50,000 use rail transit between Hachioji and Tokyo or other cities. To facilitate commuters and passers-by, Hachioji station provides rail transit and shuttle bus service, as shown in Fig. 5.18, achieving a seamless connection between rail transit and bus.

FIGURE 5.15   Soka University campus, about 5 km from Hachioji station.

FIGURE 5.16 Comprehensive building integrated with a department store in Hachioji station.

FIGURE 5.17 Commercial street around Hachioji station.

FIGURE 5.18   Eight Princes station provides a complete rail transit station and bus service connection service.

Compared with Tama New Town, Hachioji city is not a simple city. In addition to its residential function, it also undertakes part of the education and scientific research functions of the Tokyo Metropolitan Area. The entire region has business and education as its core industry, and the balance of occupation and residence is better than that of Tokyo Metropolitan Government and Tama New City.

5.6.4   Shinsan township

Shinsan township is located in Sango city, Saitama Prefecture, in the northern part of Tokyo, and about 20 km from the heart of Tokyo. It takes about an hour to get to Tokyo by rail. Like most small cities around Tokyo, a large shopping center has been built near the rail transit station, and residential areas are arranged on the periphery. Fig. 5.19 shows a large shopping center near the New Sanxiang station. Due to the limited number of jobs available in Sanxiang city, most of the residents of Xinsan township use rail transit to commute for work in Tokyo. In 2010, the ratio of daytime population to the total population of Sango City was about 85.63%[12], which was far lower than 112.1% of Tachikawa City with developed industries and 99.7% of Hachioji City[13]. However, due to the low quality of residential buildings in Shinsan township and the general community environment, as shown in Fig. 5.20, residents living here are generally from low-income groups.

FIGURE 5.19 Shopping center near Shinsango station, Saitama Prefecture, 19 km from Tokyo.

FIGURE 5.20 Shinsan township residential area around Xinsanxiang station.

5.7 Summary of the main Tokyo city characteristics and urban transportation development experience

The Tokyo Metropolitan Area centered on Tokyo city is the most populous metropolitan area in the world. The city is large and complicated, and has deep reference significance for the development of China's giant cities. In summary, Tokyo has the following experience in the development of urban and urban transportation.

5.7.1 Full practice of traffic-guided urban development

The Tokyo Metropolitan Area is a metropolitan area built on a rail transit network. The rail transit network is like a blood vessel through the metropolitan area, meeting the travel needs of the metropolitan area and supporting the development of urban activities. In the Tokyo Metropolitan Area, high-density commercial districts and mature residential communities have been built around a single rail transit site; universities with lower density or large courtyards, research institutes, and communities are connected via public transport networks and the nearest rail station. The construction of the rail transit system in Tokyo began at the end of the 19th century. Over the past 100 years, the development of the city of Tokyo has been inseparable from the construction of its transportation network. The extension of rail transit has fueled the direction of urban development. Traffic-guided urban development in Tokyo did not face intractable traffic accessibility problems at the start of construction and development. In the continuous development of the city, the rail transit can keep up with the continuous development of the city through the strong passenger transport capacity. The city and transportation promote each other and develop together, so that the world's most populous Tokyo metropolis operates efficiently.

As the world's largest commuter circle and living area, Tokyo supports modern cities with an excellent rail network, achieving an 88% green traffic rate, and a private car rate of only 11%, which is a large development for a megacity. Tokyo's urban structure is well worth considering. Large-scale long-distance commuting leads to high urban operating costs, and residents' commuting times are high. Commuting time takes up a considerable portion of people's time. From the perspective of transportation, although the urban structure of Tokyo is multicentered, the characteristics of traffic demand are essentially single-centered, with all traffic focused toward the Tokyo Metropolitan Area. The future development direction of Tokyo should be to form a truly multicenter urban structure. To this end, it is necessary to promote the balance between residence and employment, adjust the urban functional layout, and achieve the goal of shortening the commuting distance.

5.7.2 Integrated development of rail transit stations and their surrounding areas

The integrated development of rail transit stations is one of the characteristics of the development of the Tokyo rail network. In its long-term development, Tokyo has accumulated a great deal of experience that can be of value elsewhere. In the various rail stations, there are differences in the form of development and the nature of land use. Specifically, they can be divided into two types: the transportation hub of Tokyo Metropolitan Area and the common sites of other regions.

The Tokyo Metropolitan Area rail transit hub, represented by Shinjuku, has a long history of development. In terms of integrated development, the advantages outweigh the disadvantages. Relying on the huge passenger flow through the transportation hub, department stores and office buildings have been integrated with the site and have attracted a large number of people, not only facilitating the travel of residents, but also improving the efficiency of the urban operation and adding vitality to the development of the city. Its positive effects have been remarkable. However, the superior traffic conditions at the hub also significantly increase surrounding land prices, and the development strategy of the Tokyo government has been to create a commercial deputy center. The subcenter to the center of the orbital site is used for the development of high-density commercial areas. The problems of long commuting time caused by the land use purification phenomenon are also serious.

The construction of the Tokyo rail network and the surrounding land use are well integrated. For example, the Tokyo Shinjuku transportation hub has a daily passenger flow of up to 3.6 million. It has set up more than 100 rail transit entrances and exits in a transportation hub area of about 2 km^2, directly connecting with large passenger flow distribution points, forming a rail network and an underground system. The three-dimensional transportation system of rail transit and underground trails, which directly serves the surrounding land use, is the biggest feature of Tokyo's urban transportation development.

The remarkable integrated development effect of ordinary rail transit stations is represented by sites such as Tachikawa, Hachioji, and Shinsan township. Commercial facilities near the site and homes that are slightly further away naturally form a community around rail network facilities. The site provides good accessibility for the community, and the community provides a source of business for the site businesses; they promote each other and have become a good foundation for the development of Tokyo.

5.7.3 Vigorously promote the bus service with rail transit as the main travel mode

Public transportation in the Tokyo Metropolitan Area mainly relies on rail transit, which has excellent facilities, with good site coverage rate, a high departure frequency, and high punctuality. The full-time rail transit rate of

FIGURE 5.21 The underground street system in Japan is the best in the world. This image shows the Tokyo Metro Station 1st Street, Tokyo station.

FIGURE 5.22 Tokyo Yaesu underground street.

residents is 30%, and the share of public transit is at the leading level in the world. This rate can reach 48%, and the bus use rate is among the highest in the world. At the same time, the Tokyo District Department has reduced the demand for motor vehicle travel and increased the bus use rate by charging high parking fees.

5.7.4 System development of underground space resources, underground commercial, parking, and sidewalk systems are extremely important globally

The transportation hub of the Tokyo Metropolitan Area is an underground space that is fully utilized for large-scale traffic. In large transportation hubs such as Tokyo station and Shinjuku station, underground space is fully used for commercial developments (see Figs. 5.21 and 5.22). Considering the different needs of passengers at different locations in the hub, the types of shops and modes of operation of the underground commercial blocks are different, allowing passengers to purchase items conveniently and quickly.

References

[1]   Demographia. <http://www.demographia.com/> [accessed 26.09.13].

[2]   Wikipedia, Capital circle (Japan). <http://ja.wikipedia.org/wiki/%E9%A6%96%E9%83% BD%E5%9C%8F_(%E6%97%A5%E6%9C%AC)> [accessed 26.09.13].

[3]   Economic and Social Research Institute, Cabinet Office, 2010 Prefectural Economic Accounts. Tokyo; 2013.

[4]   U.S. Department of Commerce, Bureau of Economic Analysis. <http://www.bea.gov/ itable/index_error_regional.cfm> [accessed 22.10.13].

[5]   Lingling Guo. Significance of Tokyo Urban Planning for Beijing Urban Planning [D]. Beijing: University of International Business and Economics; 2005.

[6]   Ministry of Land, Infrastructure, Transport and Tourism, 2010 Metropolitan Traffic Census Metropolitan Area Report. Tokyo; 2012.

[7]   Summary of the 5th Tokyo Metropolitan Area Person Trip Survey (Traffic Survey) by the Ministry of Land, Infrastructure, Transport and Tourism. Tokyo; 2010.

[8]   Ministry of Land, Infrastructure, Transport and Tourism annual report on the metropolitan area development in 2012 (Metropolitan area white paper). Tokyo; 2013.

[9]   Yan Ying, Luo Chunxiao. Japan-Kyoto-Road Exhibition Philosophy Analysis. [J]2012;8:76.

[10] Shinjuku City Statistics. Daytime Population by Town of Chome in Shinjuku City (Estimated). <http://www.city.shinjuku.lg.jp/content/000087559.pdf> [accessed 26.09.13].

[11] Number source: Wikipedia, Tama New Town. <http://ja.wikipedia.org/wiki/%E5%A4% 9A%E6%91%A9%E3%83%8B%E3%83%A5%E3%83%BC%E3%82%BF%E3%82% A6%E3%83%B3> [accessed 26.09.13].

[12] Patmap city information. Population information of Misato-shi (Saitama): daytime, households, population density. <http://patmap.jp/CITY/11/11237/11237_MISATO_popul.html> [accessed 26.09.13].

[13] Population Statistics Division, Statistics Department, General Affairs Bureau. 2010 Census: Daytime Population in Tokyo. Tokyo; 2013.

Chapter 6

Paris, France

Chapter Outline

6.1 Introduction to Paris

Paris is the capital of France, its largest city and also the center of French politics, economy, culture and business. Paris is the fourth largest city in the world after New York, London, and Tokyo.

Eco-Cities and Green Transport. DOI: https://doi.org/10.1016/B978-0-12-821516-6.00006-0

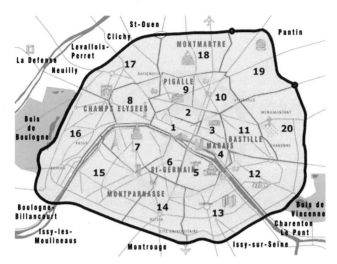

FIGURE 6.1 Schematic diagram of the administrative division of Paris. *From http://fr.hujiang. com/new/p24850/.*

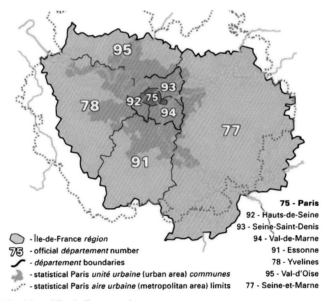

FIGURE 6.2 Map of Ile-de-France region.

Paris has 20 districts as illustrated in Fig. 6.1 shows, with a population of 2.23 million (2009) and a density of around 21,276 (persons/km^2).

Greater Paris includes the city and seven surrounding provinces, as shown in Fig. 6.2. Paris city is area no. 75, which is a small area in the center of Greater Paris. Areas 92, 93, and 94 are next to Paris city and the areas on

TABLE 6.1 General population distribution and area in different regions of Paris.

	Paris	Hauts-de-Seine	Ile-de-France
Area (km^2)	105	176	17,174
Population (millions)	2.234	1.429	12.16
Density (persons/km^2)	21,276	8137	708

the edge are nos. 77, 78, 91, and 95, which are much larger than Paris city. Greater Paris covers an area of 17,174 km^2, and has a population of 12 million. The population density is around 708 (persons/km^2).

Table 6.1 shows the area, population, and density of Paris city and Greater Paris. It can be seen that the population is not evenly distributed. The population density of Paris city is 30 times higher than that of Greater Paris, which decreases from the center to the edge of the region.

6.2 Urban structure and land use

The urban planning of Paris city has a long history, with the city spreading as the economy developed. In 1958, the La Defense was built, which created an axis extending for 8 km and brought the prosperity to the northwest of Paris. After World War II, the urban construction was accelerated as the economy developed. From the 1960s Paris started to build a new city in the suburbs to prevent the city from spreading in a disorderly fashion, developing two main axes for the city and constraining the urban space. That resulted in the city being much more balanced.

The following orders should be considered when building a satellite city. First, the satellite city cannot be too far from the main city and the transport links should be fast and convenient. Second, job—house balance and public facilities should be taken into consideration, which makes the satellite city livable and attractive. Third, the natural environment should also be paid attention to. Usually, green belts can be set between the periphery of the new town and the main city.

Manila Valley is a successful case of new town planning and construction. Manila Valley New Town is located in the eastern part of the development axis in the north of Paris. It covers an area of about 150 km^2. Rail Line A runs through the new town, providing convenient transport. Along the rail line, a series of urban developments has been carried out, thus realizing the integration of public transport and land use. At the same time, the RER line of rail transit and expressways has greatly improved the accessibility of these areas. The steps of construction give priority to residential areas, attracting the population and reserving areas for subsequent commercial development. With its rapid and orderly development, Manila Valley New Town has been recognized as a successful case in Paris in recent years.

6.3 Motorization and travel demands

The average number of daily trips per capita in Greater Paris is 3.87 (2010), of which 37.8% are car trips and 20.2% are on public transport. The corresponding number in Paris city is 4.15, with 13% being car trips and 29% on public transport.

Fig. 6.3 shows the distribution of trips within and between the districts of Greater Paris in 2007. It can be seen that the number of trips within the districts is greater than the number of cross-district trips, which also reflects the job−housing balance and the improvement of public facilities in the construction of Paris New Town.

Table 6.2 shows the distribution of travel distance between residents in Paris city and Greater Paris, as illustrated in Figs. 6.4 and 6.5, respectively. It can be seen that the travel distance within Paris is relatively short, with 61% of journeys within 1.5 km and 79% within 3 km, while the travel distance in Greater Paris is relatively long, with only 8% of journeys within 3 km and 49% over 10 km.

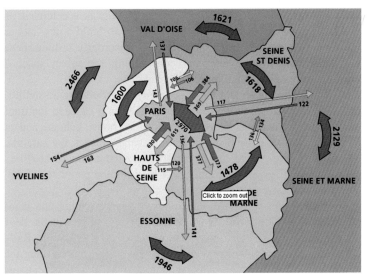

FIGURE 6.3 Number of trips within and between regions in of Greater Paris (in thousands, data in 2007). *From Paris Transportation and Travel Report (2007).*

TABLE 6.2 Distribution of travel distance between residents in Paris city and Greater Paris.

Distance (m)	≤300	300–900	900–1500	1500–3000	3000–5000	5000–10,000	10,000–20,000	≥20,000
Paris city (%)	27	23	11	18	13	8		
Great Paris (%)	2			6	11	32	32	17

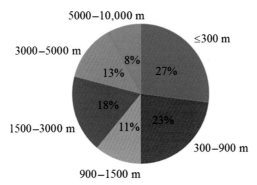

FIGURE 6.4   Travel distance distribution in Paris City. *From Paris Transportation and Travel Report (2007).*

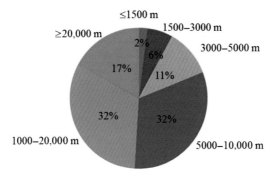

FIGURE 6.5   Travel distance distribution in Greater Paris. *From Paris Transportation and Travel Report (2007).*

In Greater Paris, there are 4.81 million motor vehicles (2010), with 80% of households owning at least one car, and the number of vehicles per thousand population is more than 395. Meanwhile in Paris city, there are 673,000 motor vehicles, with 310 motor vehicles per thousand population. Only 41.6% of families own cars, which is significantly lower than the rate in Greater Paris.

Table 6.3 and Fig. 6.6 show the mode split in Paris in 2007. It can be seen that the proportion of pedestrian travel in Paris is as high as 54%. Public transport is the main mode of motorized travel, accounting for 29%, while private cars account for only 13%. This illustrates that the transport in Paris is green and sustainable.

6.4   Public transport

Paris has a strong public transport system consisting of Metro, suburban railways, and buses. As shown in Table 6.4 and Fig. 6.7, there are 16 Metro

TABLE 6.3 Travel mode split in Paris (in 2007).

Mode	Walking	Bicycle	Public transportation	Private car	Motorcycle	Other motor vehicles	Green transport total
Mode split (%)	54	1	29	13	2	1	84

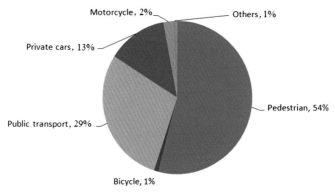

FIGURE 6.6   Travel mode split in Paris (in 2007). *From Paris Transportation and Travel Report (2007).*

TABLE 6.4 Public transport infrastructure in Paris.

Mode	Basic data
Metro	16 lines; mileage 201 km; 298 stations, average station spacing 670 m; Passenger turnover 2.643 billion person km/year
Bus	64 lines; mileage 597.2 km; 1311 buses; 1356 stations, average station spacing 440 m; passenger turnover 3 billion person km/year
Suburban railway	Five RER lines (A, B, C, D, and E), mileage 1296 km, 388 stations; average station spacing 3.34 km

lines with a total mileage of 201 km. The coverage rate of Metro stations is 100% within a radius of 500 m. The total mileage of the bus lines is almost 600 km and that of bus-only roads is 189 km. The total mileage of the suburban railway (including the five RER lines) is 1296 km. Transfers between subway, bus, suburban railway, and other public transport modes is also very convenient.

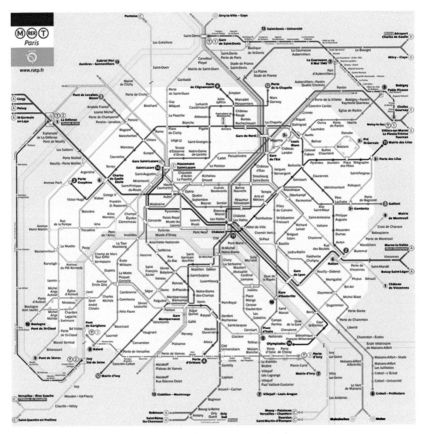

FIGURE 6.7 Map of the Paris Metro system. *From Wikipedia.*

Fig. 6.8 shows a bus-only road in Paris. The smooth road surface and clear signs provide good road conditions for safe bus operation. Fig. 6.9 is the network map of the Paris bus-only roads. The red lines in the map indicate completely isolated express bus lanes, and the blue lines indicate marked isolated bus-only roads. Paris has a wide coverage of bus lines and dense stations, which provide convenient bus travel conditions for residents.

6.5 Walking and bicycles

According to the 2007 Paris Traffic Travel Report, Paris's urban pedestrian traffic has the largest proportion among the world-class metropolises, at 54%. Paris can be described as a pedestrian city. Compared with pedestrian traffic, bicycle traffic has a small proportion of only 1%. In order to promote the development of bicycle transportation, the Paris municipal government has implemented a "bicycle city" plan since 2007, which includes the following aspects.

FIGURE 6.8 A bus-only road in Paris.

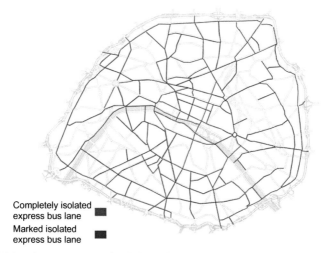

Completely isolated
express bus lane ■

Marked isolated
express bus lane ■

FIGURE 6.9 The network map of Paris bus-only roads. *From Paris Transportation and Travel Report (2007).*

6.5.1 Bicycle rental system

This system includes 206,000 rental bicycles, and the rental fee is very affordable. For long-term users, the annual rental fee is 29 euros; for short-term users, the rental fee is 1 euro per day or 5 euros per week. The system uses a self-service leasing method and the stations are interconnected, so that bicycles can be rented and returned at any station.

Paris's bicycle rental system has played a major role in promoting the use of bicycles by citizens and tourists. Paris's rental bicycles were used 2.5 million times in 2010, 3.13 million times in 2011 (Fig. 6.10), and 3.5 million times in 2012, showing a steady annual increase. It is estimated that one-third of bicycle users use the bicycle rental system. Accordingly, the share of bicycles in Paris increased from only 1% in 2007 to 3% in 2010.

6.5.2 Built bicycle lane

Since 2000, the bicycle lane mileage has been increased, reaching 400 km at the time of writing (Fig. 6.11).

Table 6.5 shows the mode split of transport in Paris in 2010. Comparing with the data in Table 6.3 for 2007, it can be seen that the public transport share increased from 29% to 31.73%, the private car share decreased from 13% to 9.91%, and the bicycle share increased from 1% to 3.11%. The walking share decreased from 54% to 52%, and the green traffic share increased by 3.15 percentage points. This shows that the idea of green transportation in Paris is becoming increasingly popular, and the car share is declining. The car share, at less than 10%, is not only rare in Western developed countries, but also compares favorably with the trend in developing countries. Hence, the experience of Paris is worth learning from.

FIGURE 6.10 Bicycle rental site in Paris.

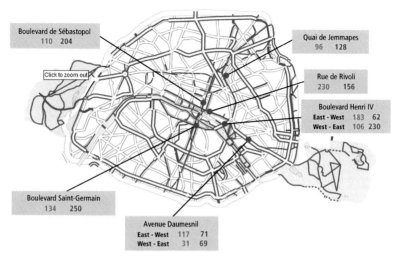

FIGURE 6.11   Bicycle traffic volume in some areas of Paris. *From Paris Transportation and Travel Report (2007) (Blue number: 3 a.m. peak volume. Red number: 3p.m. peak period).*

TABLE 6.5 Transport mode split in Paris (2010).

Mode	Walking	Bicycle	Public transportation	Private car	Motorcycle	Other motor vehicles	Green transport total
Mode split (%)	52	3.11	31.73	9.91	2.14	0.80	87.15

6.6   Case study

6.6.1   La Defense transfer hub

La Defense transfer hub, as shown in Fig. 6.12, is a comprehensive transport hub that integrates rail transit (high-speed railway, Metro line), highway, urban public transport, and other modes of transport. La Defense transfer hub was planned and constructed began in 1985, and was built in the mid-1990s. At the same time, the development of the surrounding area was started. The number of passengers now taking Metro Line 1 at La Defense transfer hub has reached 17.54 million per year, with 74,000 passengers per working day, with 29.72 million passengers taking RER-A at a rate of 120,000 passengers per working day. With other modes of transportation, there is a total of 400,000 people transferring here daily.

FIGURE 6.12 La Defense integrated transportation hub. *From http://www.visions-de-paris.com/.*

La Defense transportation hub is a three-dimensional, multimodal integrated hub, consisting of Metro Line 1, tram line T2, RER-A regional railway, and an underground suburban railway line. The powerful railway system connects La Defense closely with the central district of Paris. Ground levels 1−3 are the main roads for motor vehicles, avoiding plane crossing through overpasses. Parking lots are located on both ground and underground levels, providing more than 26,000 parking spaces. Pedestrian roofs are built on platforms of 3−5 floors. As shown in Fig. 6.13, pedestrian roofs consist of slopes, pools, greening, shops, sculptures, squares, and other comfortable pedestrian spaces, with a total area of 67 ha. Pedestrians and vehicle traffic are completely separated by the pedestrian roof, which provides a safe walking environment for pedestrians, and also greatly facilitates pedestrian connection between buildings.

La Defense is worth learning from not only for the transportation hub itself, but also from the integrated development and utilization of the transportation hub and surrounding land. Houses, schools, office buildings, and commercial centers are built around the transportation hub, so that the whole area has comprehensive urban functions. La Defense has become the deputy center of Paris.

The La Defense covers a total area of 750 ha, with 250 ha developed in the first phase (2004−09).

The area plans to have 185,000 jobs, 25,000 residents, 7 colleges, and 45,000 students. However, in the first phase of the project, 80% of the land was used for office buildings, while the residential and commercial areas were very small. As shown in Table 6.6, the land use pattern is not mixed

FIGURE 6.13   Pedestrian roofs of La Defense.

TABLE 6.6 Planning area of the phase I project of La Defense (2004–09).

	Business	Park	Total
Phase I construction area (Ha)	160	90	250
Office building area (10,000 m^2)	215	32	247
Number of residential units (10,000 units)	1.01	0.56	1.57
Residential capacity (10,000 people)	2.1	1.83	3.93

and the job—house ratio is unbalanced. The area lacks vitality and interaction with residents during nonworking hours. Therefore, in the second phase of the project, it was stipulated that the area of office buildings could not exceed 50%. The second phase of the project covered an area of 320 ha, and was completed around 2015.

6.6.2   Ecological community

In recent years, Paris has promoted the construction of ecological communities, the goal of which is to save resources and energy, maintain a good ecological environment, and provide convenient, safe, healthy, and comfortable ecological communities for citizens. In order to achieve this goal, ecological communities follow the principles of "3D," that is, density, diversity, and design.

Density. With the increase in population density and the level of urbanization, land resources are very valuable. Therefore a city must maintain a high density of development in order to save valuable land resources and provide public infrastructure and transport.

Diversity. In order to reduce the traffic demand of travelers, such as commuting, school, shopping, and medical treatment, the land use should be mixed and job—house balance and public service should be taken into consideration. This will reduce the use of private cars, and promote green transport such as walking and bicycle use. The types of buildings should also be diversified, including residential buildings, office buildings, schools, and hospitals.

Design. In order to create a comfortable, beautiful, warm, and livable community environment, it is necessary to carefully design buildings, streets, water systems, parks, landscapes, etc.

The trapezoidal block community located in the southwestern suburb of Paris and near the terminal of Metro Line 9 is a successful case which embodies the principles of "3D" and the development and construction of an ecological community. It has the following characteristics.

1. High density and good design

   The trapezoidal block covers an area of 40 ha, with an average floor area ratio (FAR) of 3.5—4.0. The density is relatively high and the land use is relatively compact. However, there is no sense of high density and boring buildings. As shown in Fig. 6.14, the buildings are hidden in the green trees. Red buildings and green plants together form a

FIGURE 6.14 Buildings among green trees.

FIGURE 6.15   Wide public square design along with tall buildings.

bright and vibrant community landscape. At the same time, architectures use different heights, shapes, and colors of buildings to avoid creating a boring environment. As shown in Fig. 6.15, the buildings
distributed around the street parks have their own characteristics. Even
if buildings are the same, different facade styles and colors are
designed at different levels. These varied architectural designs make
the community not only meet the requirements of high-density development, but also create a colorful architectural landscape. In addition,
residents pay attention to the decoration of the building itself, using the
balcony and facade of the buildings to increase the green area and
improve the overall environment and landscape of the community as
shown in Fig. 6.16.

2. Mixed land use

The community has residential buildings, office buildings, schools,
kindergartens, commercial facilities, and other functional buildings. As
shown in Fig. 6.17, office buildings and residential buildings are in the
same block, less than 100 m apart. The bottom floor of residential buildings contains shops, enabling residents to work and shop nearby and conveniently. There are also schools in the community, as shown in
Fig. 6.18, so that students can go to school nearby, which is safe and convenient. Because the community takes into account the job−house balance and basic needs of the community, the cross-district travel demands
and car use are reduced greatly, which is positive for saving energy and
protecting the environment.

FIGURE 6.16   The greening of balconies and roofs integrates the greening system in space.

FIGURE 6.17   Office buildings and residential buildings are in the same block.

3. Walking and bicycling

The pedestrian roads in the community are not accessible to motor vehicles. The two sides of the pedestrian roads are fully greened, as shown in Fig. 6.19. The pedestrian roads are not only safe and peaceful walking spaces in the community, but also provide links between buildings and open spaces. As shown in Fig. 6.20, the green landscape extends to every building along the pedestrian roads. Residents can easily enjoy the green ecological

FIGURE 6.18 Schools in the community enable students to travel short distances to school.

FIGURE 6.19 Fully greened pedestrian roads.

environment. There are also bicycle roads with clear signs, as shown in Fig. 6.21. Pedestrian roads and bicycle lanes provide a good environment for green traffic in the community and greatly reduce the use of motor vehicles, which plays a significant role in residents' health and the ecological environment, increasing the opportunities for residents to communicate, and creating a harmonious and healthy community atmosphere.

FIGURE 6.20   The green landscape extends to all buildings along the pedestrian roads.

FIGURE 6.21   Bicycle roads with clear signs.

6.6.3   The left bank of Paris

The left bank of Paris is located in the 13th District of southern Paris, along the Seine river, and was originally a rail area. The famous Austerlitz Railway Station is located here. The area covers an area of about 130 ha, with a population of about 15,000 inhabitants and about 5000 jobs.

Urban planning for this area started in 1979 and has been ongoing for about 40 years, and remains in place. Every building also employs design competition on the basis of planning. On the other hand, a model exhibition hall is to be constructed to give the public the opportunity to understand the planning to enable public participation.

The upper layer of the railway area is covered with a building, with the area divided into three layers. The top houses the main building, the center is the original railway (the surface layer), and the underground is the parking area. In this way, not only is the bare landscape of the original railway area in the city properly covered, but also a large amount of land for development and construction has been released, providing funds for the project.

6.7   Successful experiences

Paris, a famous European and international city, is very large and has a high population density in the central city. It can be used as a very good example for the construction and development of large cities in China. Paris's old city has a history going back hundreds of years. In the process of development, great attention has been paid to the protection of the historical features of the city. The construction of the new city mainly relies on the rail transit network to develop the new city. The job—house balance is also paid attention to, and has been a great success. Paris has the following major experiences in urban and transport development.

1. Mixed land use and job—house balance

Paris pays great attention to mixed land use, whether for the whole city or specific areas. In the construction of new towns, special attention should be paid to the provision of jobs in order to prevent the phenomenon of a "bed city." For example, before 2009, the first stage of the La Defense Project overemphasized the construction of commercial buildings and office buildings, resulting in a job—house imbalance. The second stage of the project was adjusted over time. Having a good job—house balance and mixed land use can reduce the travel distance of residents and promote green transportation, such as walking and bicycle use.

From the travel distance structure in Paris, it can be seen that within 105 km^2 of Paris, 79% of the travel distance is less than 3 km and 92% of the travel distance is less than 5 km. The travel structure of Paris is determined by the urban structure and land use pattern.

2. Priority to public transport and rail transit

The public transport system of Paris is well organized and covered. The average distance between Metro stations is only 670 m, and the average distance between bus stations is 440 m, which one of the highest densities in the world. The public transport share is 31.73%, with the green traffic share being 87.15%, and the private car share only 9.95%. No other city in the world has achieved such successful green traffic system results.

Several new towns in Paris have been built mainly around rail transit stations, which not only facilitates the connections between new towns and central urban areas, but also promotes the development of rail transit.

3. Promotion of bicycle use

Paris is promoting the development of bicycle traffic through the implementation of a "bicycle plan." For example, by building bicycle roads and establishing bicycle rental system, the share of bicycles in Paris has increased from 1% in 2007 to 3% in 2010.

4. Ecological community construction

Good design can avoid the negative impact of high density. High density is necessary for high-impact land use, and has a positive effect on the promotion of public transport. In the case of sustainable communities in Paris, high-density development is adopted. In order to avoid the negative impact of high-density development on the landscape, the community has completed a fine green space design, while the buildings have also been designed carefully, reducing the oppressiveness of high buildings (Figs. 6.22–6.24).

5. Make full use of underground and ground space

Paris's urban construction is very good at making full use of space. For instance, the pedestrian roof in La Defense separates people from vehicles and enables people to walk to any building in the area. In the

FIGURE 6.22 Buildings surrounded by greenery.

FIGURE 6.23   Beautiful architecture and ecological harmony.

renovation of the railway area in southern Paris, a large number of buildings were built above the railway, making full use of the empty space.
Meanwhile, it also conceals the unattractive landscape of the original
railway, and has replaced it with an emerging architecture and urban
landscape.

6. Implement the urban plan, and promote public participation

Most of Paris's planning began in the middle of the 20th century, and
has been continuously optimized in the following decades, thus guaran-
teeing the harmony of the overall urban style and construction ideas. In
the process of urban planning, Paris also attaches great importance to
public participation. Through a planning exhibition, the plan is shown to
the public. It also designs many models to display the future of the city.
At the same time, the opinions of the public can be acquired to improve
the plan.

FIGURE 6.24 The landscape cannot only be viewed, but also embraced.

Further reading

Paris Transportation and Travel Report; 2007.

Parisrivegauche. <http://www.parisrivegauche.com/>.

Suping B. Interpreting the spatial morphology of Paris. South Architecture 2010;(1):74−6 [in Chinese].

Tao L, Yuemin N. The experience and Enlightenment of the development of new cities in international metropolises. China Urban Studies 2013;8(1):31−41 [in Chinese].

Wikipedia: Paris. <http://zh.wikipedia.org/zh-cn/>.

Xiaoshan Z. Characteristics and Enlightenment of traffic planning and management in the CBD La Defense of Paris. China Academic J Electron Publishing House 2012;26(4):64−6 [in Chinese].

Chapter 7

Seoul, Korea

Chapter Outline

Eco-Cities and Green Transport. DOI: https://doi.org/10.1016/B978-0-12-821516-6.00007-2

7.1 Overview of the city

Seoul is the capital of the Korea, and the political, economic, cultural, and commercial center of Korea. Seoul Metropolitan Area is the second most heavily populated metropolitan area in the world after Tokyo Metropolitan Area.

Seoul Metropolitan Area consists of three parts: Seoul special city (Seoul city), Incheon metropolitan city, and Gyeonggi Province. In 2012, Seoul Metropolitan Area covered an area of 11,818 km², accounting for 11.8% of the total area of Korea [1], with a population of 24.546 million, containing 49.3% of the total Korean population and with a population density of 2079 people per square kilometer. Within this area, Seoul city covers 605 km², accounting for only 0.6% of the total area of the Korea, with a population of 10.3 million, representing 20% of the total Korean population [2], and a population density of 17,024 people per square kilometer. The data shown in Table 7.1 compare the area, population, and population density of Seoul city and Seoul Metropolitan Area, and Fig. 7.1 is a map of Seoul city.

7.2 Urban structure and land use

Seoul is located in a basin surrounded by mountains that are more than 500 m high, and the city has been an important natural fortress since ancient times. Seoul has an obvious single-center urban structure. Its main roads network consists of 3 loop lines and 19 radiation lines, and the city functions are concentrated. In 1962, Korea began to implement an Urban Planning Law, and in

TABLE 7.1 Population, land area, and population density of Seoul metropolitan area.

	Seoul Metropolitan Area	Seoul city
Land area (km^2)	11,818	605
Population (thousands)	2456.4	1030
Population density (people/km^2)	2079	17,024

FIGURE 7.1   A map of Seoul city.

1966, the master plan for Seoul was introduced. In 1970, in order to promote the development of the area to the south of the Han River, this master plan was revised. In 1972, Seoul formulated a 10-year plan for urban development with the aim of creating a world-class city [3].

The Han River runs across Seoul from east to west in a "W" shape. The original urban development in Seoul was mainly around Mapo and Hannan ferry crossing along the river. With more bridges across the Han River, land transportation has gradually improved, and the urban development of Seoul has gradually expanded to both the north and south, forming a series of residential areas such as Jamsil. The areas of Myeong-dong and Dongdaemun to the north of the Han River are concentrated in traditional Korean-style streets and administrative offices; however, areas such as Gangnam to the south of the Han River were developed during the high-speed economic growth period after the Korean War and became the working and living areas for new high-income classes.

The development of Seoul is closely linked to the Han River. The history, culture, and development of the city are all centered around the Han River and its tributaries. Seoul is now paying more attention to the harmonious coexistence of people and nature. The Cheonggyecheon restoration project, the development of the Han River tributary riverside area, and the construction of the leisure area along the Zhonglangchuan River are examples illustrating the concept of better ecological protection.

7.3 Motorization and traffic demand characteristics

In 2012, there were 20.01 million trips in Seoul city daily, and the number of trips in Seoul Metropolitan Area was 49.66 million [4]. According to Seoul's 2006 traffic survey data, the total number of motor vehicles entering and leaving Seoul city daily is about 3.53 million, including 1.47 million vehicles entering and exiting the central city area [5].

In 2012, the daily average one-way commute time for Seoul residents was 42.5 minutes. The average one-way commute time for commuters in Gyeonggi-do and Incheon in the Seoul Metropolitan Area was 60.5 and 71.2 minutes [4]. Residents in the suburbs have longer commute times.

Seoul's motor vehicle number initially grew rapidly, however, the growth rate then decreased. This reflects a deepening of the understanding of green transportation and effective policy-oriented results in Seoul. In 1970, there were only 60,000 vehicles in Seoul, but in the 1980s the number of motor vehicles increased rapidly, and by 1998 it had reached 2.2 million. Since the start of the 21st century, the number of motor vehicles in Seoul has been growing slowly. In 2006, the number of motor vehicles was 2.857 million [5], with the total number in 2012 at 2.97million [6]; this equates to 288 vehicles/1000 people.

Fig. 7.2 shows the changes in the number of motor vehicles in Seoul, Seoul Metropolitan Area, and Korea from 1980 to 2011. It can be seen from the figure that compared with the overall growth of motor vehicles in Korea, the growth in Seoul was slower.

Vehicle ownership (1000)

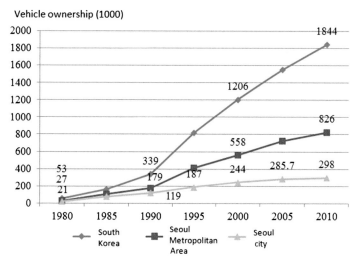

FIGURE 7.2 Motor vehicles quantity in Korea, Seoul Metropolitan Area, and Seoul city. *Data from e-National Indicators 2011; JaeHak OH. Korea's challenge for people-centered mobility; 2013.*

7.4 Public transportation

Seoul has a developed public transportation system, and the share of subway and bus use being more than 60%. In 2012, Seoul subway had a total of nine lines, which formed a network throughout the city (Fig. 7.3). The total length of the lines is 327.1 km, with 190 km being underground. There are 302 stations, with the average station spacing at 1.08 km. The average daily passenger number is 6.9 million and the annual number is 2.5 billion [2]. Table 7.2 lists the basic operational data of each subway line in Seoul in 2012.

In 2004, Seoul city carried out a comprehensive reform of the bus service. The density, speed, comfort, and punctuality of the bus service were greatly improved, attracting more citizens to bus travel in the process. The share rate for bus travel reached 27.8%, making it the second largest method of travel in Seoul. Table 7.3 shows the rates of use of different modes of travel. It can be seen that the share of green transportation has reached 67.9%.

7.5 Construction of bicycle and pedestrian transportation systems

7.5.1 Change of motor vehicle roads back to walking and bicycle spaces

In the early 1990s, Seoul began to enter a period of rapid growth in motorization, and the proportion of private car travels rose rapidly. At that time,

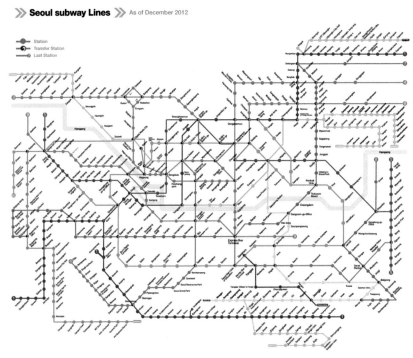

FIGURE 7.3 Seoul subway map. *From Seoul Metropolitan Rapid Transit to the World (2012).*

TABLE 7.2 Operation of subway lines in Seoul (2012).

Line no.	Length (km)	Passenger number/day (thousands)	Line load (thousand passengers/ km · day)	Station number	Average station spacing (km)
Line 1	9.9	450	45.5	10	0.99
Line 2	60.2	2005	33.3	50	1.20
Line 3	38.2	755	19.8	34	1.12
Line 4	31.6	832	26.3	26	1.22
Line 5	52.3	812	15.5	51	1.03
Line 6	35.1	487	13.9	38	0.92
Line 7	46.9	864	18.4	42	1.12
Line 8	17.7	231	13.1	17	1.04
Line 9	27	266	9.9	25	1.08
Total	318.9	6702	21.0	293	1.09

TABLE 7.3 The shares of different modes of transportation in Seoul.

Mode	Subway	Bus	Walking/bicycle/others	Personal vehicle	Taxi
Share (%)	35.2	27.8	4.9	25.9	6.2
Green transport (%)	67.9				

Source: Data from Seoul Metropolitan Rapid Transit to the World (2012).

the government adopted measures to build more infrastructure in response to the rapid growth of motor vehicles. Urban roads and transportation resources were gradually occupied by motor vehicles. The space reserved for bicycles and pedestrians gradually decreased. However, traffic congestion in Seoul has become more and more problematic.

In the 21st century, Seoul has gradually realized that a return of green transportation with bicycles and walking is the core method to achieving sustainable development. This is a trend that modern people are increasingly turning to. Based on this concept, the city again began to construct bicycle roads. The city squares previously occupied by motor vehicle roads were returned to walking and bicycle roads, giving citizens a safe, healthy, and energy-saving green passage space. Significant changes were made between concept and practice. The set of photographs in Fig. 7.4 records the historical evolution of the space in front of the City Hall from 1974 to 2013. At first, the square was completely a pedestrian space. Then from the 1970s to the 1990s, the motorway gradually occupied the square during the rapid mobilization trend. Finally, it returned to a green square in the 21st century [7].

The changes to Diving Bridge over the Han River is another example emphasizing the greater importance paid to pedestrians and bicycles. In 1999, the bridge was a two-way four-lane road bridge that was solely for motor vehicles (Fig. 7.5A); in 2013, half the road was converted into a bicycle and pedestrian way, and the roads were color-paved, as shown in Figs. 7.5B and C. Both motor vehicles and pedestrians have a continuous and safe passage space. Green transport is not only a planning concept, but is also implemented in the final plans.

7.5.2 Bicycle rental system to serve the "last kilometer"

In addition to road space, in order to encourage more people to use bicycles, Seoul has set up free bicycle rental points near subway stations and bus stops to solve the "last kilometer" problem that people need to walk to the final destination after getting off the train or bus, which can be inconvenient. As shown in Fig. 7.6, travelers can rent bicycles at subway stations or bus stops.

FIGURE 7.4 Historical changes to the square in front of Seoul City Hall. ① In the 1940s, the Square was completely pedestrianized; ② in 1952, vehicles passed by Sejong Avenue in front of City Hall; ③ in the 1960s, the pedestrian space was compressed on both sides by the road; ④ in the 1980s, the pedestrian space was further compressed; ⑤ in 1999, the intersection in front of the City Hall forbade left turns to reduce congestion; ⑥ and ⑦ in 2013, the square in front of City Hall was re-established, allowing the return of pedestrians after the greening reforms. *From Liren D. A report on the urban traffic from congestion to smooth in Seoul, South Korea; 2013.*

7.5.3 Building a safe, comfortable, and warm pedestrian transportation system

In order to create a safe and comfortable walking environment, Seoul has also carried out a lot of detailed work on the construction of its pedestrian system. Fig. 7.7 shows a pedestrian walkway at the intersection in the city center. Considering the large pedestrian flow, in order to improve efficiency, clear directional signs are set out on the walkway. There is also a safety island so that pedestrians are able to cross the road safely and in an orderly fashion. Fig. 7.8 shows an isolation guardrail between the pedestrian road and the motor vehicle lane on a relatively narrow road, with a speed limit sign setting the maximum speed for motor vehicles at 30 km/h. The sculptures set on the side of the pedestrian road shown in Fig. 7.9 create a welcoming walking space for people to enjoy and relax.

FIGURE 7.5 The changes to Diving Bridge passing over the Han River (1999–2013). (A) Two-way four-lane road solely for motor vehicles in 1999; (B) half of the road was converted into a bicycle and pedestrian way in 2013; (C) the south entrance of the bridge in 2013.

FIGURE 7.6 Bicycle rental station near a subway station.

FIGURE 7.7 Directional signs on a pedestrian walkway.

In some traditional historical districts there are pedestrian-only streets. Figs. 7.10 and 7.11 show the famous Korean traditional art district Insadong. It is a pedestrian-only space with a rich historical and artistic atmosphere. People can enjoy Korean art, shopping, and leisure in a peaceful and comfortable environment.

FIGURE 7.8   Guardrails to separate pedestrians from motor vehicles.

FIGURE 7.9   A relaxing space on a pedestrian road.

The modern Seoul city center is full of high-rise buildings with a wide range of road types but limited area. In order to build and improve the pedestrian transportation system, Seoul also attaches great importance to the development and utilization of underground space. Connections between subway stations and pedestrian underground passages serve as underground shopping streets, making the underground passages spacious, bright, and warm in winter and cool in summer, offering shelter from inclement weather.

FIGURE 7.10 Entrance to the Insadong traditional art district.

Underground shopping streets have greatly improved the underground space environment and conditions, and constitute an important part of the walking system, as shown in Figs. 7.12 and 7.13.

7.6 Building a green transportation system: reform of the Seoul bus system

Its single-center urban structure and high-density demographic characteristics mean that Seoul faces enormous transportation needs. With the continuous suburbanization of residential land, Seoul has faced increasingly serious traffic problems since the 1980s. The total number of daily trips has been rising and the distance per trip has grown. This has resulted in tremendous pressure on urban traffic. From 1970 to 2002, the total number of trips from downtown Seoul to the suburbs increased from 5.7 to 29.6 million, a fivefold

FIGURE 7.11   Streets in the Insadong traditional art district.

increase. In this serious situation, effectively increasing the share of public transportation and reducing the use of private cars in the central city is the best solution. However, with the opening of the Seoul Metro in 1974, the share of bus travel began to decline. By the end of the 20th century, the average speed of buses was as low as 19 km/h due to the absence of bus-only lanes. Half of the buses used diesel, resulting in seriously polluted exhaust fumes. The bus on-time rate was low, and transfer to subway stations was not convenient. A series of problems have seriously reduced the attractiveness of the bus system. In the decade from the late 1980s to the end of the 1990s, the share of bus travels dropped by almost 50%, and bus companies suffered serious economic losses [8]. Facing these severe conditions, Seoul needed to conduct a comprehensive bus reform, adjust bus operation strategies, improve bus service levels, improve bus system facilities, and increase the attractiveness of bus use.

FIGURE 7.12 Subway station connection area serving as a shopping street and walking space.

FIGURE 7.13 Pedestrian underground passage serving as a shopping street and walking space.

In August 2003, Seoul established a Bus System Reform Citizen Committee and established a joint seminar covering government, nongovernmental organizations, bus companies, and experts. The Committee and seminar discussed reform measures including bus network layout, the charging model, and the operating system. On July 1, 2004, Seoul's comprehensive

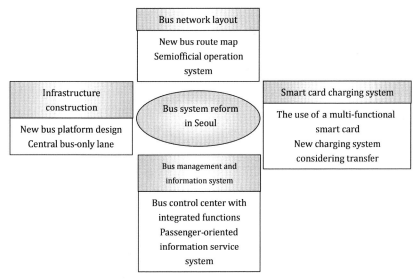

FIGURE 7.14 Main measures for bus system reform in Seoul.

bus reform was officially implemented. As shown in Fig. 7.14, bus reform was mainly implemented in the bus network layout, infrastructure construction, a smart card charging system, bus management, and an information system relying on the Internet and GPS positioning.

7.6.1   Reforming of the bus network layout

Seoul organizes and integrates bus lines, and divides bus lines into red, blue, green, and yellow colors. The red bus line is a suburban line connecting the far-reaching satellite city and the central city (Fig. 7.15); the blue bus line is the backbone bus line connecting regional centers in Seoul and runs through the main roads (Fig. 7.16); the green bus line connects the community and the surrounding subway stations and large bus stops to facilitate transfer of community residents; the yellow bus line is the loop line that works as a supplement to the network. Through the above layout reform, bus operation has been optimized. At the same time, the bus reform divides Seoul into several areas with numbering of each of the areas. The bus line number is associated with the area number to help passengers understand the route. By optimizing bus routes, the density of bus stations in the built-up areas of Seoul has increased from 14.6 per square kilometer to 16.35. The number of bus lines has increased by 94, as shown in Table 7.4.

FIGURE 7.15 The red bus line connects the far-reaching satellite city and the central city.

FIGURE 7.16 Blue buses running on the main road in Gangnam district.

7.6.2 Infrastructure construction

In order to improve the speed and ensure the smooth traffic flow of buses, 14 central bus lanes with a total length of 177.6 km were built, as shown in Fig. 7.17. These central bus lanes form the backbone of the Seoul BRT system. Through the high-quality island-style bus station and bus priority signal system, the bus speed has been significantly improved, and the average speed

TABLE 7.4 Bus lines before and after the bus system reform.

Before the reform (July 2003−December 2003)		After the reform (July 2004−December 2004)	
Type	Number of lines	Type	Number of lines
City	253	Downtown main line (blue)	90
Advanced	37	Branch line (green)	328
Rapid	12	Suburban rapid line (red)	39
Circle	66	Circular line (yellow)	5
Total	368	Total	462

Source: Data from Jingzhe J. The road to sustainable public transport − experience and achievements in Seoul's bus reform; 2006.

FIGURE 7.17   Bus lanes in the middle of the road.

has increased from 13 km/h before the reform to 17.3 km/h. At the same time, the buses were upgraded and more than 300 low-floor natural-gas buses were purchased for the operation of the red suburban line and the blue main line, to improve passenger comfort (Table 7.5).

7.6.3   Smart card charging system

The transportation system data for Seoul were unified, and the T-Money smart card launched, which joined the previously separate charging systems

TABLE 7.5 Vehicle speeds on main roads.

Main road	Length (km)	Car speed (km/h) (June 4, 2004)	Bus speed (km/h) (December 4, 2004)
Dobong-Mia	15.8	11.0	22.0
Susak-Songsan	6.8	13.1	21.5
Kangnamdero	5.9	13.0	17.3

Source: Data from Jingzhe J. The road to sustainable public transport — experience and achievements in Seoul's bus reform; 2006.

for subways and buses, making it more convenient for people to travel. After the reform, the charging system set transfers within half an hour as a different component of a trip, and does not charge for these separately, reducing travel expenses and improving the utilization rate of buses. At the same time, the T-Money smart card entitles the holder to a discount, and can also be used in shopping malls and convenience stores, enhancing its appeal.

7.6.4 Public transportation management and information system construction

The bus control center can integrate the road data detected by the intelligent transportation system and the bus position data obtained from GPS to control the operation of the bus system in a timely fashion. Through mobile phones and other facilities, the system can also provide passengers with real-time bus information to reduce their waiting time.

Through the above-mentioned bus system reform measures, Seoul's bus service level has been significantly improved, and the number of passengers using buses and the proportion of bus travel are also increasing. In the first year after the reform, the number of bus trips increased by almost 10%, the bus accident rate dropped by 10.66% [9], and the punctuality rate increased by 10% [5]. After the reform, the bus service level was greatly improved. Fig. 7.18 shows the average daily passenger volume change of Seoul Metro and buses from 2000 to 2010 (excluding special tickets such as 1-day passes and 3-day passes). There is a significant increase in the daily passenger trips using buses since the reform in 2005.

7.7 Case of an ecological city construction: Cheonggyecheon restoration project

Cheonggyecheon is located in the center of Seoul, and is a stream from Beiyue and Nanshan mountains, passing through the city from west to east.

Daily trips / 10,000 trips

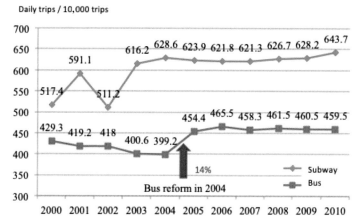

FIGURE 7.18   Average daily trip changes in the Seoul subway and bus systems. *Data from JaeHak OH. Korea's challenge for people-centered mobility; 2013.*

In ancient times, Cheonggyecheon was one of the main areas where Seoul citizens lived and worked. However, after the 1900s, many farmers who sold farmland flocked to Seoul and became the "urban poor," living on both sides of Cheonggyecheon. As the urban poor increased, the living environment along Cheonggyecheon progressively deteriorated [10]. In addition, environmental problems such as flooding and river pollution during the dry season became increasingly serious.

In the Japanese occupation period from 1937 to 1942, in order to reduce the pollution, Seoul began to cover the river. Due to shortage of funds, the project was not entirely completed. In 1958, after its economic recovery, the Cheonggyecheon Coverage Project was restarted and began to fully cover the remaining 5.6-km section without any environmental work. In the 1970s, Korea's economy developed rapidly and the number of motor vehicles began to increase exponentially. In order to solve the traffic problem in Seoul, the Seoul metropolitan government built a two-way four-lane elevated highway above Cheonggyecheon. With the rapid development of the Korean economy, many residential and commercial centers were built along the Cheonggyecheon expressway, which quickly become the bustling center of Seoul.

As Cheonggyecheon crosses Seoul's busiest city center, Cheonggyecheon expressway has become Seoul's main channel for east−west traffic, and this traffic has increased rapidly. However, problems such as traffic noise, exhaust emissions, and urban isolation have begun to plague the development of the surrounding area. At the start of the 21st century, Cheonggyecheon was almost dry, and the its concrete cover was also in poor condition. Cheonggyecheon, as an important area in the center of Seoul, has caused serious negative effects on Seoul's image.

FIGURE 7.19 Cheonggyecheon restoration project started in 2003. *From Liren D. A report on the urban traffic from congestion to smooth in Seoul, South Korea; 2013.*

In order to improve the city's quality and reproduce the beautiful scenery of Cheonggyecheon, in July 2003, the Cheonggyecheon restoration project was launched. This project mainly includes destruction of the Cheonggyecheon elevated road, ecological restoration, and waterfront ecological cultural landscape construction, as shown in Fig. 7.19. In October 2005, after more than 2 years' construction, the Cheonggyecheon Waterfront Park replaced the original elevated road and reappeared in full public view. The Cheonggyecheon restoration project not only returned the history and culture that Cheonggyecheon had lost for more than 30 years, but also played a positive role in enhancing Seoul's urban image and strengthening Seoul's economic strength. Figs. 7.20 and 7.21 are the scenes before and after the restoration project, respectively.

After the renovation, Cheonggyecheon was restored its original appearance, and the clear stream and beautiful waterfront environment returned to Seoul, greatly improving the ecological and landscape environment around the stream and improving the quality of urban residents' lives. Cheonggyecheon's water is mainly derived from the Han River, and is supplemented by groundwater and rainfall. When necessary, the reclaimed water is used, so that Cheonggyecheon is not cut off all year round. The daily water volume from the Han River to Cheonggyecheon is about 120,000 tonnes. The river landscape varies from west to east, depending on the function of the city. The western part of the river is located in the city center and adjacent to the national government buildings. It is an important political, financial, and cultural center. The banks of the river are paved with granite slabs to form a

FIGURE 7.20 Before the restoration project, a highway on the river, with the river was underneath.

FIGURE 7.21 After the Cheonggyecheon restoration project, the appearance of Cheonggyecheon was restored.

FIGURE 7.22 Hydrophilic platform at the western part of Cheonggyecheon.

hydrophilic platform, as shown in Fig. 7.22. The central section of the river passes through the famous Dongdaemun Market, which is a famous small commodity wholesale market in Korea. It is a communal area for ordinary citizens and tourists. Therefore the design of this section emphasizes the leisure characteristics of the waterfront space, focusing on a classical and natural scene. Stone and grass-covered slopes are built into the south bank of the river, and the continuous hydrophilic platform is set on the north bank, as shown in Fig. 7.23. The eastern section contains a residential area and a mixed residential and commercial area, therefore the design reflects the natural ecology. The main feature is the step-by-step cross-river channel that gives people the feeling of a return to nature [11]. Green natural slopes on both sides of the river help to maintain the natural ecology of the river, as shown in Fig. 7.24. The restoration project has a significant improvement effect on the surrounding environment. The open and clear river constitutes a wind channel and greenway in the central city of Seoul, which greatly improves the surrounding air quality and reduces summer temperatures. With regard to the average temperature near Cheonggyecheon, before the renovation, it was 5°C higher than other parts of Seoul city. Since the renovation, the temperature is 3.6°C lower than other regions. In addition, there are many sculptures with ethnic characteristics on both sides of Cheonggyecheon. During traditional festivals, Seoul citizens also hold a national-style celebration at Cheonggyecheon, as shown in Fig. 7.25. Cheonggyecheon not only has a beautiful natural environment, but has also become a place to show Korean traditional culture and ethnic customs.

FIGURE 7.23   Hydrophilic platform in the middle section of Cheonggyecheon with stone and plant revetment.

FIGURE 7.24   Revetment covered by natural plants and a step-by-step crossing in the eastern section of Cheonggyecheon.

Before the implementation of the Cheonggyecheon renovation project, there was concern that the renovation project would have a significant adverse impact on the traffic in downtown Seoul, however, this has not happened. On the contrary, in the process of the project, by promoting public

FIGURE 7.25   Cultural elements with ethnic characteristics at Cheonggyecheon.

FIGURE 7.26   Comfortable pedestrian walkway near Cheonggyecheon.

transportation, the traffic pressure in the city center has been greatly allevi-
ated. As shown in Fig. 7.26, a comfortable pedestrian walkway has been
built along the river. Long-distance travel uses subways or buses to keep
people away from the hustle and bustle experienced in the past. The project
has found a way to restore the ecological environment and improve public
transport development.

7.8 Experience from Seoul's green transportation and eco-city development

Represented by the bus system reform and the Cheonggyecheon restoration project, Seoul has undertaken great efforts to become an eco-city with green transportation. It is worth learning from the exploration and innovation carried out in Seoul.

7.8.1 Respecting nature, paying attention to the city's taste, and the return to beautiful Cheonggyecheon

From the construction of the elevated road above Cheonggyecheon during the period of high-speed economic growth in the past, to spending the huge sum of US$360 million to remove the highway and return it to its natural ecology and beautiful river landscape, Seoul has learnt precious lessons on ecological civilization development. In the Cheonggyecheon restoration project, a perfect hydrophilic environment and cultural atmosphere has been carefully designed, creating a fresh and natural landscape along the river in the heart of a modern metropolis. The positive effects, including city-level improvement, better quality of living, and the promotion of tourism are difficult to measure financially. This is a major change to the concept of development, which embodies the idea that in the process of moving toward a modern society, economic development is not all that is needed. Coordinated development with the natural environment is also very important.

7.8.2 Facing the rapid motorization trend, changing the concept, and decisively implementing public transport priorities

As the capital of Korea, and with a population of more than 10 million, Seoul has faced problems in transportation development. Since the 1970s, the large-scale construction of roads, represented by the Cheonggyecheon expressway, has been attempting to catch up with the growing number of motor vehicles in the city. However, it turns out that the expansion of road networks alone has not been able to meet the endless space requirements of cars.

Facing a background of a deteriorating environment and increasingly prominent traffic problems, Seoul has recognized that it is impossible to meet endless growing demand from motor vehicles by building roads. It is also unacceptable to use limited land resources to enable more motor vehicles to be used in the city. The only way to serve traffic needs in a megacity is to mainly rely on public transportation.

Therefore at the start of the 21st century, Seoul decisively promoted bus system reform, effectively improved the service quality of the bus system by optimizing bus routes, set up bus-only lanes, and improved the intelligent management level of public transportation. In this way, the share of public transportation use has reached 63%. Urban traffic congestion has eased, laying a strong foundation for the healthy development of the city.

7.8.3 Creating a safe and comfortable bicycle and pedestrian transportation system

During the period of rapid economic development, Seoul attempted to expand the road space of motor vehicles and improve the traffic capacity of motor vehicles by squeezing bicycle and pedestrian passages. However, road traffic became increasingly crowded. From these lessons, Seoul has firmly established a public transport priority, and given the precious passage space back to bicycles and pedestrians. The reconfiguration of the passage space in front of Seoul City Hall and the case of Diving Bridge are examples. The changes from this concept in Seoul and the series of traffic reforms under the guidance of the green transportation concept represent the development direction of the city. China is currently facing rapid urbanization and motorization, and could learn a valuable lesson from the example of Seoul.

References

[1] National statistics: population and family: resident population and area statistics. <http://kostat.go.kr/portal/english/surveyOutlines/1/1/index.static> [accessed 14.12.13].
[2] Seoul Metropolitan Government. Seoul Metropolitan rapid transit to the World. Seoul; 2012.
[3] Jun W. Lessons from the adjustment of the national administrative center of South Korea and Russia. Outlook Weekl 2012;28:56−7 [in Chinese].
[4] JaeHak OH. Korea's challenge for people-centered mobility. KOTI; 2013. (Lecture, Singapore, October, 2013).
[5] Jiyun J. A comparative study of urban transportation system between Beijing and Seoul, 49. Beijing Jiaotong University; 2009 [in Chinese].
[6] Status quo of automobile registration December 2012. <http://www.index.go.kr/egams/stts/jsp/potal/stts/PO_STTS_IdxMain.jsp?idx_cd = 1257> [accessed 5.10.13].
[7] Liren D. A report on the urban traffic from congestion to smooth in Seoul, South Korea; 2013 [in Chinese].
[8] Jingzhe J. The road to sustainable public transport - experience and achievements in Seoul's bus reform. Urban Transp China 2006;4(3):27−32 [in Chinese].
[9] Kim KK, Kim GG. KOTI knowledge sharing report: Korea's best practices in the transportation sector issue 1: bus system reform in Korea. Gyeonggi-do: The Korea Transport Institute; 2012.
[10] Hong L, Qing Y. The restoration and reconstruction of Cheonggyecheon in Seoul of South Korea. Int Urban Plan 2007;22(4):43−7 [in Chinese].
[11] Jun W. Enlightenment of ecological reconstruction of Cheonggyecheon in South Korea for China. China Dev 2009;9(3):15−18 [in Chinese].

Chapter 8

New York, United States

Chapter Outline

8.1 Introduction

New York is a global city, in which the headquarters of the United Nations are located. New York Metropolitan Area is one of the largest metropolitan areas in the world. It exerts tremendous global influence on business and finance, directly affecting the global economy, finance, media, politics, education, entertainment, and fashion.

New York City consists of five districts: Manhattan, Queens, Brooklyn, Bronx, and Staten Island, with a total area of 783.83 km^2, a population of 8.3367 million (2012) and a population density of 10,636 person/km^2. The metropolitan area of New York covers a total area of 30,670 km^2, with a population of 22.214 million (2012) and a population density of 724 people per square kilometer (Table 8.1).

8.2 Urban structure and land use

New York's urban divisions comprise different groups of people with diverse urban life patterns. Building function, architectural form, and density, however, are not very different in the different regions.

In 1968, the New York Association for Regional Planning (NYRP) suggested five principles for the second metropolitan plan to stop urban sprawl:

1. Establish new urban centers, provide high-level public utilities, and transform New York into a multicenter metropolis;
2. Revise the zoning policy of new housing to provide more diverse communities;
3. Raise the level of service facilities in the old city as far as possible, improve the environment, and reattract people of all income levels;

Eco-Cities and Green Transport. DOI: https://doi.org/10.1016/B978-0-12-821516-6.00008-4

173

TABLE 8.1 New York City population profile (in 2012).

	Metropolitan area of New York	New York City	Manhattan
Area (km^2)	30,670	783.83	87.46
Population (million)	22.214 (2012)	8.3367 (2012)	1.619
Density (person/km^2)	724	10,636	18,511

4. New urban development should keep the main parts of the region in a natural state;
5. Create a suitable public transport system.

Therefore the development of urban land in New York has been transformed, focusing on improving the density and efficiency rather than urban expansion and sprawl. According to a statistical analysis, from 1988 to 2002, the area of construction land in New York City increased from 579.77 km^2 to 622.16 km^2, with a net increase of 42.39 km^2 in 14 years and an average annual increase of 3.03 km^2. From 2002 to 2006, the area of construction land in New York City decreased by 2.90 km^2. From 1988 to 2006, the per capita construction land of New York was reduced from 79 square meters per person to 75 square meters per person, and the land use efficiency and density have been greatly improved.

8.3 Motorization and travel demands

In New York City, more than half (54.5%) of households do not have cars. There are about 195,000 motor vehicles, with 234 motor vehicles per 1000 people. In Manhattan district, 77% of households do not have cars. However, in New York Metropolitan Area, there are more than 8 million motor vehicles and about 360 motor vehicles per 1000 people.

New York City residents travel 3.46 times a day (in 2009), 3.53 times a day on weekdays and 3.28 times on weekend. Travel distance per capita is 8.32 km. Walking and public transport are the main modes of travel (as shown in Fig. 8.1; Table 8.2). Public transport, especially rail transit, is clearly dominant for commuting, with a share of 28.7% (Fig. 8.2).

8.4 Public transportation

The New York Metro is the backbone of public transport and one of the most complex and long-standing underground railway systems in the world. At

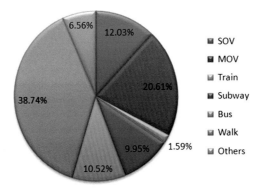

FIGURE 8.1   Mode split of commuter transportation modes for New York residents.

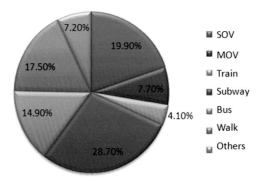

FIGURE 8.2   Mode split of commuter transportation modes for New York residents. *From New York State 2009 NHTS Comparison Report.*

TABLE 8.2 Mode split of all-purpose transportation modes for New York residents.

Mode	Walk	Train	Metro	Bus	Multiple occupancy vehicle (MOV)	Single occupancy vehicle (SOV)	Others	Green transport in total
Mode split (%)	38.7	1.6	10.0	10.5	20.6	12.0	06.6	067.8

present, there are 469 stations and a combined line length of 1056 km. It connects four of New York's five districts. Fig. 8.3 shows the entrance to the Metro station in Manhattan. Fig. 8.4 shows a map of the New York Metro line. Staten Island is connected to Manhattan and Brooklyn by ferry and bus services.

FIGURE 8.3 Metro station entrance in Manhattan.

New York City has 3700 buses on 200 routes in its five districts. There are 14,000 bus stops in the city. Bus lanes and tow zones are part of the design of Manhattan Central Business District (CBD). In some sections, no vehicles are allowed to turn "left" or "right" except buses. "No-stop zones" are set up in the main streets of the city center. Bus stops are available at all intersections and streets of the CBD. Buses are inexpensive, convenient for transfers, and on time, and are the most commonly used means of transportation for New Yorkers. In 1993, New York State passed the State's Business Tax Act and imposed taxes on the purchase of gasoline, which were used to subsidize the development of urban public transport.

8.5 Walking and bicycle use

Michael Bloomberg, the mayor of New York, said, "New York has a completely new public bicycle rental system that people don't have to pay for." The plan was officially launched on May 27, 2013, with 300 sites and 6000 bicycles, and expanded to 600 sites and 10,000 bicycles in the next few years. By the start-up date, 15,000 people had joined the public bicycle rental program.

Users of the New York public bicycle rental system can choose to pay an annual fee of $95 and rent bicycles indefinitely for no more than 45 minutes each time during the year, or buy daily tickets of about $10 and weekly tickets of $25 from June 2, 2013, but the rental time for both tickets could not

FIGURE 8.4   Map of New York Metro. *From http://www.yoyv.com/oci_subwaymap/3624.html.*

exceed 30 minutes. Rental time is based on the time that the bicycle is returned to the site. If this time is exceeded, the consumer pays extra fees.

The transportation department is also considering bicycle lanes to connect Manhattan. This dedicated road will include a "bicycle road" along the East River and Hudson River, and will be connected with the "green bicycle line" in the North and South Bronx, Brooklyn, and Queens. The transportation department would also improve the sidewalks in Lower Town, where the sidewalks are in a state of disrepair and very inconvenient for pedestrians.

8.6 Traffic management

In the 20th century, in response to a large increase in the number of motor vehicles, governments at all levels in New York invested a lot of money in the construction of the highway system, gradually replacing the railways. After the end of World War II, in 1956 and 1958, the government further amended the Federal Highway Grants Act, which emphasized the dependence on motor vehicles and stimulated the enthusiasm of residents' car purchases. During this period, by investing heavily in the construction of transportation infrastructure, New York's economic level was significantly improved, began to enter a rapid motorization development stage, and completed the basic construction of all urban road networks.

During that period, New York was overly focused on economic growth and urbanization, which led to an increase in motor vehicle ownership that outpaced road construction. Highway construction further stimulated the desire to buy motor vehicles, thus resulting in a chaotic cycle of motorization. Because of the lack of correct understanding and management determination, New York traffic entered a very difficult period in the 1970s. Traffic congestion occurred in different regions, greatly affecting residents' ability to travel. Road construction led to land development chaos, and urban air and environmental quality also significantly declined.

After receiving criticism and suggestions from many sides, New York's traffic development strategy has changed significantly. New York has made major adjustments to the city's traffic policy, implemented the public transport priority development policy, and increased investment in public transport. Urban transport authorities have invested a total of $48 billion in improving and upgrading public transport, including 356 routes and subway stations, train replacement, and bus replacement. The punctuality rate of the Long Island Railway has increased from 85% to 93.2%, and punctuality rate of railway from the metropolitan to northern has increased from 80.5% to 97.5%. In terms of improving the utilization of public transport systems and supporting economic growth, 400,000 jobs were created in the central business districts. From 1992 to 2000, investment in railways increased by 28%, and investment in buses and Metros increased by 44%.

In addition, New York has a peak-time charging policy. Two tunnels and four overpasses between New York and New Jersey charge differently depending on the intersection and time. The charges are highest when traffic congestion is most severe and public transport is abundant. They are relatively low when public transport is crowded and there are few alternatives. There is also a discount for travel at night (midnight to 6 a.m.) and other off-peak times.

Inside the city, New York has relieved traffic congestion by setting up a large number of one-way lines. New York's financial, commercial, and entertainment zones have long been formed and are extremely difficult to adjust. Many roads are narrow and parking lots are inadequate. Under these circumstance, the streets of New York are mostly one-way, not only for its four-lane broad avenues, but also narrow streets, which guarantee the overall smooth flow with partial restrictions. New York's street layout is mainly grid-like. The two adjacent one-way lines generally have different directions. Two-way lines are set up on the main streets every few streets to provide a variety of choices for pedestrians. New York also relies on the continued construction of parking lots around suburban Metro stations to reduce private car access to Manhattan's Central Business District, facilitating residents' use of the intercity rail system and parking.

8.7 Experiences

New York is not only the largest city in the United States, but also a globally famous metropolis. The main experiences in the development of New York City and transportation are as follows:

1. Attention should be paid to urban mixed land use, job—housing balance, and public service facilities

    The most important feature of New York City is mixed land use. Although Manhattan is the most famous financial center in the world, there is still a large amount of land for residential use. The bottom levels of residential buildings integrate business and residence. This also leads to short travel distances in New York City (8.3 km), with 38.7% of daily travel being walking, which greatly reduces the number of motor vehicle trips in the city, thus effectively alleviating traffic congestion.

2. Adjust transportation policy in time and promote public transport

    After the lessons of the chaotic development of the car infrastructure, the New York City Government adjusted its transportation development strategy and policy in time. Public transport priority is the most prominent feature of New York. Private cars in urban areas are generally not as convenient as public transport. High station coverage and punctuality make public transport services very competitive. A series of financial measures also ensure the sustainable development of public transport.

3. Control of car use in the central area

    In the city center, New York gives priority to public transport on one hand, and controls car use by various means on the other hand. High cost and inconvenient parking, strict turning requirements, and severe penalties have greatly limited the use of cars in the city center. New York's public transport share leads among American cities thanks to these measures.

4. Vigorously promote walking and bicycle use

    According to New York Plan, by 2030, the proportion of bicycle lanes to all roads in New York will be increased from 1:15 to 1:10, encouraging citizens to use bicycles. At present, New York's bicycle lanes have reached a total distance of about 1000 km, making New York the most effective city in the United States in promoting bicycle travel. A public bicycle system is also under construction. Walking and bicycle use are the most ecological modes of transportation. Guaranteeing the traffic space for walking and bicycle will effectively prevent cities from car dependence.

5. Implement a scientific traffic management policy

New York has achieved good results in alleviating traffic congestion using various means of traffic management. New York has divided the city into grids for planning and predicting possible traffic conditions over the next 30 years. It has identified the public transport priority as its greatest strategic requirement, and established the most developed public transport system in the United States. Car use and parking are strictly managed in the city center. At the same time, an intelligent traffic management system has been established to improve the efficiency of traffic operation.

Further reading

Bus first. New York City Bus. China.com <http://www.china.com.cn/> [in Chinese].
<http://www.census.gov/>.
Lu J, Song J. Analysis of urban land use structure and its zoning characteristics in New York City. Res Urban Develop 2010;(12):90−7 [in Chinese].
New York State 2009 National Household Travel Survey (NHTS) Comparison Report.
PlaNYC. A Greener, Greater New York; 2010.
Shi Y, Huang Y. The characteristics of New York City Planning and Its Enlightenment to Shanghai. World Reg Stud 2010;19(1):20−7 [in Chinese].
Wikipedia. New York. <http://zh.m.wikipedia.org/>.

Chapter 9

London, United Kingdom

Chapter Outline

9.1 City profile

London is the capital of the United Kingdom of Great Britain and Northern Ireland, and also the largest city and financial center in Europe. London ranks with New York, France, Paris, and Tokyo as one of the five world-class cities. Since the 18th century, London has been one of the most important political, economic, cultural, artistic, and entertainment centers in the world [1,2].

9.1.1 Geographical conditions

London is located on the western part of the European continent, on the plain of southeast England, facing Paris, France, across the English Channel. The Thames flows from Windsor Castle on the west of London into the city and continues to flow 88 km east to the river mouth in the North Sea. The Thames and its dense tributaries give London a developed water system, the Blue Ribbon Network, formed by a water system with a plentiful water supply, convenient urban water transport channel, and diverse ecological environments and animal habitats. The river forms beautiful natural scenery and a unique human landscape (as shown in Figs. 9.1 and 9.2).

FIGURE 9.1 Landscape along the Thames river.

FIGURE 9.2 Another image of the landscape along the Thames river.

9.1.2 Administrative divisions

"London" usually refers to Greater London, including the city of London and 32 other boroughs [3]. The London administrative divisions are shown in Table 9.1. The city of London is at the center, consisting of the City of London, most of the City of Westminster, Camden, Hackney, Islington, Lambeth, Southwark, Tower Hamlets, and parts of Wandsworth, Kensington, and Chelsea; the 13 urban areas around the city of London are called inner London, and the outermost 19 urban areas are called outer London.

9.1.3 Population characteristics

According to the regional population statistics of London in 2011 [4], the population and population density of different areas, such as Greater London, outer London, and inner London, are as shown in Table 9.2. The area of London is 27 km^2, and the population density of inner London is nearly twice that of the entire Greater London area.

TABLE 9.1 Administrative districts of London.

Number	Region name	Number	Region name
0	London	17	Merton
1	City of London	18	Sutton
2	City of Westminster	19	Croydon
3	Kensington and Chelsea	20	Bromley
4	Hammersmith and Fulham	21	Lewisham
5	Wandsworth	22	Greenwich
6	Lambeth	23	Bexley
7	Southwark	24	Havering
8	Tower Hamlets	25	Barking and Dagenham
9	Hackney	26	Redbridge
10	Islington	27	Newham
11	Camden	28	Waltham Forest
12	Brent	29	Haringey
13	Ealing	30	Enfield
14	Hounslow	31	Barnet
15	Richmond upon Thames	32	Harrow
16	Kingston upon Thames	33	Hillingdon

TABLE 9.2 London area populations and population densities (2011).

Region	Inner London	Outer London	Greater London
Square (km^2)	321	1259	1580
Population (10,000)	323	494	817
Population density (person/km^2)	10,062	3923	5171

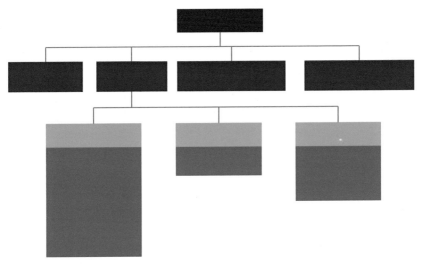

FIGURE 9.3   London traffic management system.

9.2   Urban traffic development policy

9.2.1   London traffic management system

The London traffic management system is illustrated in Fig. 9.3. Transport for London (TFL) is in the charge of the Mayor of London as a functional department of government. Transport for London includes three aspects:

1. Surface transport, including London buses, London streets, Victoria Coach Station, River service, Dial-a-Ride, and the Public Transport Office;
2. London Underground;
3. London Rail, including Docklands Light Railway (DLR), London Overground, and London Trams.

In 2009, Greater London introduced the Mayor's Transport Strategy, the objectives of which included:

1. Supporting economic development and population growth;
2. Improving the quality of life for all Londoners;
3. Improving security for all Londoners;
4. Improving transport for all Londoners;
5. Reducing the impact of traffic on climate and enhancing climate resilience;
6. Supporting the London 2012 Olympic Games and Paralympic Games.

9.2.2   London's public transport policy

Over the years, the transport department has taken active strategies and measures to control traffic demand and developed purposeful restrictions on traffic volume. The London government has planned the layout of the city's residential, work, entertainment, and other infrastructures in a rational way, enabling citizens to make full use of public transport and taking various measures to restrict the use of cars. The public transport policy of London includes the following aspects [3,5−7]:

1. Safeguarding the transportation budget
   TFL's budget covers transport operating costs, capital projects, debt servicing, and contingency costs.
2. Construction of a convenient transfer system
   When new lines are added in London, the old stations should be rebuilt or reconstructed, and transfers between buses and subways should be improved by installing elevators and appropriate adding signage.
3. Setting bus priority
   Setting up bus lanes or bus priority roads to ensure unobstructed road space for buses.
4. Improving the efficiency and level of bus services
   The main strategies include developing the reform plan for the bus system, updating the buses, and improving the service level. To achieve a wide range of data sharing, drivers are made aware of traffic status in real time, with electronic stop showing the vehicle arrival times. Traffic information services help people to make plan trip times and routes accurately, helping to save travel time and improve the attractiveness of public transportation. Advanced electronic technology has been adopted to improve the service level of the ticketing system. By introducing an express card, bus pass, and single zone pass fare, more citizens are encouraged to use public transport.
5. Extending bus running times
   Bus services between residential areas and the city center usually operate until midnight, with some continuing to operating after that time.

6. Vigorously promote pedestrian and bicycle transportation

More efforts should be made to build cycling roads in London, to create a more green and eco-friendly cycling environment, and to increase bicycle parking facilities near schools and stations.

9.2.3 London's traffic management policy

The key points of London's traffic management policy [3,5−7] are as follows:

1. Automation and modernization of traffic management

Microwave detectors are installed on highways and main arteries, and video detectors are installed on main roads and intersections, which can monitor the real-time traffic state, feeding back this information to the traffic control center so as to facilitate timely issuance of traffic instructions. The installation of a vehicle identification system on main roads, and wireless transmitters or GPS positioning equipment in buses can exchange data with data centers to realize vehicle−road communication and ensure the priority of bus signals. Crosswalks with traffic lights are equipped with devices and sensors for manually changing traffic lights on both sides to ensure the safety of crossing pedestrians.

2. Restrictions on cars

In order to achieve the priority of public transport, traveling time, road sections, and speed of cars are strictly restricted.

3. Parking management

Adequate parking spaces are set up around subway stations, railway stations, and bus terminals to provide convenience for "Park and Ride" travel and to reduce roadside parking.

4. Congestion charging

In February 2003, London began to implement the policy of congestion charging, which levied congestion charges on vehicles entering the downtown area. Different from one-time "tolls," London charged according to the areas entered, also known as the "regional permit" system, as shown in Fig. 9.4 [5].

9.3 Urban structure and land use

From the administrative area of London, it can be seen that London expands from the center of the most prosperous area: from the City of London to inner London and finally outer London. After the World War II, with the recovery of industry, London's population increased dramatically. To reduce the pressure from population and traffic in the central city, a number of satellite cities have been built at distances of 20−30 miles from the city of London. In the satellite cities, the metropolitan and regional centers are built to distribute the pressure from the downtown areas and form the Greater London of today.

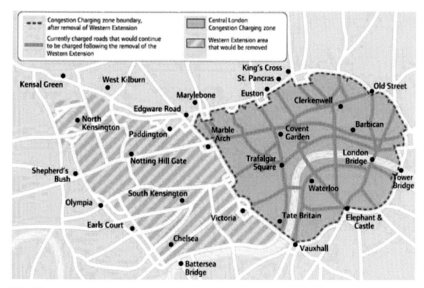

FIGURE 9.4 London congestion zone map. *From Keegan M. Mayor's transport strategy. London: Transport for London; 2009.*

TABLE 9.3 A comparative analysis of population and employment in London by region (2007).

Region	Square (km²)	Population (10,000 people)	Population density (10,000 people/km²)	Jobs (10,000)	Employment position/ resident population
City of London	27	30	1.11	120	4
Inner London	294	270	0.92	150	0.56
Outer London	1259	460	0.37	200	0.43
Greater London (total)	1580	760	0.48	470	0.62

According to statistics from 2007 [5], an analysis and comparison of the London population and jobs in different areas are shown in Table 9.3. From these data, it can be concluded that the resident population of London city is 300,000 people, offering 1.2 million jobs, which is four times that of the resident population, resulting in a large number of residents from inner and outer London who work in the city of London. Therefore the commuter traffic volume between the city and the outer areas is enormous. Tidal commuter traffic flows are evident during morning and evening rush hours.

More residents live in inner London, where the population density is nearly twice that of the overall population density of Greater London, as there are plenty of high-rise residential buildings, with a high plot ratio. The resident population was 2.7 million in 2007, with 1.5 million jobs, creating a ratio between jobs and the resident population of 0.56:1. The development intensity of outer London is lower, with a relatively small population density. In 2007, the resident population was 4.6 million, with 2 million jobs, and the ratio of jobs to resident population was 0.43:1.

9.4   Characteristics of urban traffic demands

9.4.1   Share characteristics of transportation modes

According to survey data from 2011 [8], in 2011, the average number of daily trips made by the resident population in London was 2.55 times/day, and the proportion of travel modes is shown in Fig. 9.5. Shopping and leisure were the main trip purposes, accounting for 29% and 28%, respectively, while traveling to work accounted for 23%. In 2011, the total number of daily trips in London was 25.5 million. The share rates of different modes of transportation are shown in Fig. 9.6 and Table 9.4, and the share rate of public buses (including trams) is seen to be 21.6%. The share rate of the subway was 10.6%, with 9.2% for rail, 20.6% for walking, and 33.5% for private

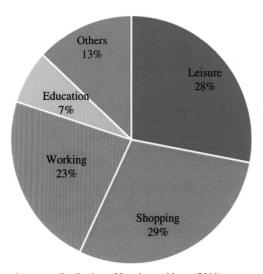

FIGURE 9.5   Travel purpose distribution of London residents (2011).

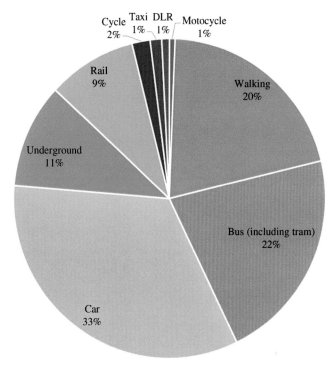

FIGURE 9.6   Share of transportation modes of London (2011).

cars. The share rate of green transportation, including public transportation, walking, and cycling, was 64.7%.

The total number of daily trips in London has increased by 11.3% compared with 10 years ago. Under the influence of a series of public transportation promotions, the shares of transportation modes have also undergone great changes: railway travel increased by 41.9%, bus travel increased by 59.7%, and bicycle travel increased by 66.6%. In relative terms, private car trips decreased by 13.0% [8].

9.4.2   Motorized characteristics

As shown in Fig. 9.7 [8], private car ownership and utilization rate in London have decreased slowly in recent years. Private car ownership was 0.779 cars per household in 2005 and 0.718 cars per household in 2011. The average number of private cars used per day was 1.054 in 2005 and 0.936 in 2011.

The main reasons for the decline in car ownership and use frequency are:

1. Development of public transport systems;
2. Increase in the population;

TABLE 9.4 Share of transportation modes of London (2011).

Transport mode	Walking	Cycling	Under ground	Rail	Bus (including tram)	DLR	Car	Motorcycle	Taxi	Total of green transport
Share (%)	20.60	1.90	10.60	9.20	21.60	0.80	33.50	0.60	1.20	64.70

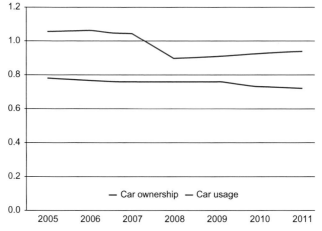

FIGURE 9.7   Changes to car ownership and use over time.

3. Increase in the plot ratio and population density in inner London;
4. Increase in the number of single families;
5. Implementation of various measures to encourage a reduction in car ownership.

It can also be seen from the graph that there was a significant decline in the use of private cars in London in 2008, as London's economy was also greatly affected under the background of the global financial crisis.

9.5   London's public transport system

London's public transport is a very advanced, powerful multidimensional public transport system consisting of subways, overground rail systems, buses, water transport, etc.

9.5.1   Underground

On January 10, 1863, London opened the world's first underground railway (London Underground). London metropolitan subway initiated the human use of underground space for transportation. At present, there are 11 London Underground lines [9], with a total length of 402 km. There are 270 stations, and the average distance between each station is about 1.49 km. Fig. 9.8 shows the London underground network [10], the service covers the downtown area and most areas north of the Thames.

The statistics for the annual traffic volume of London Underground during 2001–11 showed that over these 10 years, the passenger volume of the railway increased year on year. In 2011, London Underground carried

FIGURE 9.8 London Underground network. *From Transport for London. Standard Tube map.* *http://www.tfl.gov.uk.*

TABLE 9.5 Statistics of London Underground operating indicators (2008–12).

Year	2008	2009	2010	2011	2012
Annual mileage (100 million kilometers)	0.706	0.694	0.689	0.724	0.76
Punctuality rate (%)	96.4	96.6	95.6	97	97.6
Delay time (minute)	6.6	6.4	56.5	5.8	5.3
Degree of passenger satisfaction (score)	79	79	79	80	80

1171 million passengers annually, with a total distance of 9.519 billion passenger kilometers, with an average travel distance of 8.1 km. The average load intensity of the lines is 7981 person/(km · day).

Table 9.5 shows the annual running mileage, punctuality, delay times, and passenger satisfaction [11]. As can be seen from this table, over 5 years, the overall trend of annual mileage of London Underground has increased, and the service level has been continuously improved. By 2012, the annual mileage increased to 76 million kilometers, the punctuality rate increased to 97.6%, and the delay time decreased to 5.4 minutes. The level of passenger satisfaction scored 83.

FIGURE 9.9   Upper and lower passages at a London Underground station during rush hour.

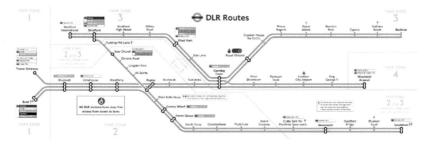

FIGURE 9.10   London Docklands Light Railway. *From Transport for London. Docklands Light Railway System map. From http://www.tfl.gov.uk.*

Fig. 9.9 shows the upper and lower passages at a London Underground station during rush hour. The subway is the preferred mode of transportation for commuters due to its high punctuality rate and the advantages of not being affected by weather and road traffic conditions.

9.5.2   The ground rail transit system

The ground rail transit system is mainly located around the south bank of the river Thames in London, with the departure interval time in the central region during peak hours at 2.5 minutes. In the suburbs, the departure interval time is 2−8 minutes. The ground rail transit system is composed of the DLR, Overground, National Railway, and London Trams.

9.5.2.1   Docklands Light Railway

DLR started running in 1987, and has a total of 15 stations and 11 operation vehicles. Today, there are seven Docklands Light Railway lines. Fig. 9.10 shows a map of the London Docklands Light Railway [12]. The service scope of the London Docklands Light Railway includes south and southeast London, connecting Bank, Beckton, Lewisham, London City Airport, Woolwich Arsenal, etc.

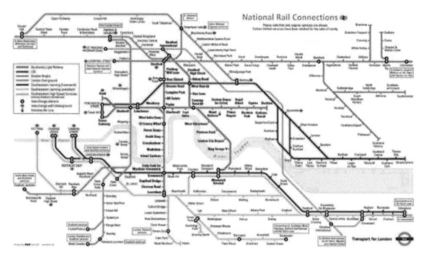

FIGURE 9.11   London Docklands Light Railway connections to other rail transit systems. *From Transport for London. Docklands Light Railway mainline rail connections map. From http://www.tfl.gov.uk.*

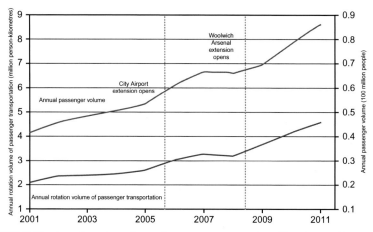

FIGURE 9.12   Annual rotation volume of passenger transportation of London Docklands Light Railway.

London Docklands Light Railway has good transfer connections with other types of rail transit, as shown in Fig. 9.11 [13]. The blue line in this figure is the Docklands Light Railway, and the black circle is the transfer station location. As can be seen from Fig. 9.12, the passenger volume of London Docklands Light Railway in 10 years has increased year on year. In 2011, the annual passenger volume was 86 million, and the total passenger distance traveled was 456 million person·km, with an average travel distance of 5.3 km. The average line load intensity is 6930 persons/(km·day).

9.5.2.2 Overground

In 2007, the overground railway (Overground) was officially put into service; 30% of Londoners can reach a station within 14-minute walk. Today, there are 6 lines and 84 stations in the London Overground system, with a total length of 86 km, and the average distance between stations is about 1.1 km. Fig. 9.13 shows the London Overground railway network [14], which serves the areas outside central London, running across the 22 boroughs of Greater London, Three Rivers of Hertfordshire, and Watford.

The statistical results of the annual traffic volume of the London Overground railway system from 2008 to 2011 [8] show that in those 4 years, the number of passengers on the London Overground increased year on year. In 2011, the number of passengers on the London Overground was 103 million, and the total distance traveled was 645 million person-kilometers, with an average travel distance of 6.3 km. The average load intensity of the line was 3281 persons/(km · day).

FIGURE 9.13 London Overground network. *From Transport for London. London Overground network map. From http://www.tfl.gov.uk.*

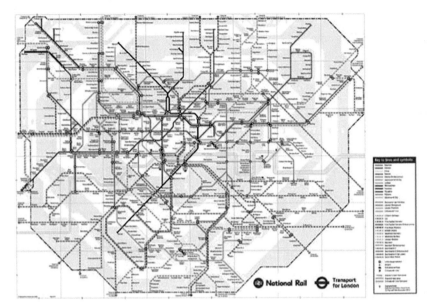

FIGURE 9.14   National Rail network in London. *From Transport for London. London Rail and Tube services map. From http://www.tfl.gov.uk.*

9.5.2.3   National Rail

The National Rail network in London is shown in Fig. 9.14 [15].

The statistical results of the annual traffic volume of London National Rail during the period from 2001 to 2011 [8] shows that in these 10 years, the overall passenger traffic showed an increasing trend. In 2011, the number of passengers on London National Rail reached 994 million, with a total distance traveled of 26.5 billion person · km, and an average travel distance of 26.7 km.

9.5.2.4   London Trams

London Trams was revived in 2000. Currently, it has four lines and 39 stations, with a total length of 28 km and an average distance between stations of 0.7 km. Fig. 9.15 shows the London Trams network [16], which mainly serves Croydon, the surrounding area, and Wimbledon. Croydon is located on the south edge of London, and jobs are concentrated in the central area, which the tram line provides access to.

Fig. 9.15 shows the statistical results of the annual traffic volume of London Trams during the period from 2001 to 2011 [8]. In these 10 years, the overall passenger traffic of London Trams had an increasing trend. In 2011, the number of passengers on London Trams reached 29.4 million, with

FIGURE 9.15 London Trams network. *From Transport for London. Tramlink user guide. From http://www.tfl.gov.uk.*

a total distance of 148 billion person·km, and an average travel distance of 5.1 km. The average load intensity of the line is 2837 persons/(km·day).

9.5.3 Buses

Buses are an important part of the London transport system; London has 700 bus routes. As early as 1897, London became one of the first cities to provide a bus service, and the London bus network is now one of the most developed public transport systems in the world and Europe's largest bus network. Figs. 9.16 and 9.17 show the red double-decker buses running through the streets of London, which have become a unique landmark of London's street landscape.

From 2001 to 2011, the statistical results of the annual bus traffic in London [8] showed that, in these 10 years, the passenger traffic of London buses increased year on year. In 2011, the annual passenger traffic of London buses was 2.344 billion person, and the weekly passenger traffic distance was 8.219 billion person·km.

9.5.4 Waterway system

The port of London is the busiest port in Britain, with its waterway and port facilities mainly concentrated in London, along the Thames River. The waterway transport system is mainly composed of the Water Bus and River Bus. The River Bus serves only the public transportation along the Thames river and is part of the public transportation system, while the Water Bus, operated by London Waterbus Company, provides other supplementary waterway public transport [3,5].

FIGURE 9.16 Buses in Oxford street, London's high street.

FIGURE 9.17   Buses in Oxford street.

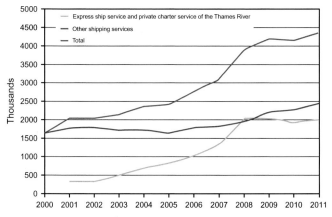

FIGURE 9.18   Change to annual passenger volume by water transportation provided by the Thames by year.

The statistical results of the water transportation service provided on the Thames in London from 2000 to 2011 are shown in Fig. 9.18 [10]. The blue line indicates the total amount, the green line represents the express ship service and private charter service of the Thames, and the red line is other shipping services. As shown in the figure, the passenger volume of water transportation has increased year on year. By December 2011, the waterway passenger volume reached 4.357 million people in London, with the passenger volume of the express ship service and private charter service of the

FIGURE 9.19 Large number of tourists cruising the Thames.

FIGURE 9.20 Sightseeing cruises along the Thames.

Thames at 1.95 million, and the passenger volume of other shipping services at 2.407 million. Sightseeing along the Thames is an extremely important tourist activity in London, attracting tens of thousands of tourists from all over the world daily. As shown in Figs. 9.19 and 9.20, sightseeing cruise ships undertake this huge sightseeing project.

9.6   Walking and cycling traffic

In 2005, London introduced a guide to improve the pedestrian system (Guide for Walking Improvement of 2005) [17], aiming to improve the service level of the walking system. The guide pointed out that the pedestrian network and ancillary facilities should meet five requirements, namely connection, happiness, striking, comfort, and convenience. In 2009, the London mayor proposed Mayor's Transport Strategy [5], including the following policies relevant to walking system:

1. Building a safe, comfortable, and attractive walking environment;
2. Making it easier for walkers to make plans for hiking trips;
3. Ensuring that walking is healthy and more accessible to the natural environment.

The London Plan issued in 2011 [3] made policies for the walking system, including:

1. Improving the construction of strategic pedestrian routes and other pedestrian routes of London;
2. Ensuring high accessibility, safety, and comfort of pedestrian routes connecting town centers and other important traffic nodes;
3. Putting forward the Legible London Initiative to provide convenient directions for pedestrians;
4. Establishment of a regulatory review mechanism to ensure that the infrastructure of the pedestrian system played its due role; and
5. Encouraging the construction of high-quality neighborhood environments and establishing the Shared Space Principles, including simple design of urban street landscape, clean pedestrian environment, and unlimited access to space.

9.6.1   Bicycle transportation system

The bicycle transportation system is an important part of the London transportation system. The London government currently encourages citizens to use bicycles. In the next 10 years, the London government will increase investment, initially with the construction of a large-scale bicycle infrastructure and an information service system to ensure safe and comfortable traffic space for cycling. Cycling will make a great contribution to environmental protection, congestion relief, and public health.

The use of bicycles has different characteristics in different areas of London. London cycling usage during 2001—11 [8] is shown in Fig. 9.21, the red line represents bicycle usage in London, the yellow line represents bicycle usage in inner London, the green line represents bicycle usage in outer London, and the blue line shows bicycle usage along the Thames.

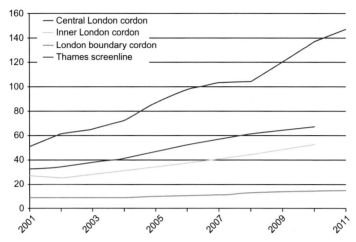

FIGURE 9.21 Variation trend in bicycle usage in London.

As a commercial and financial center, the city of London has a high building density, a large number of jobs, limited road space, large traffic demand, and serious traffic congestion. The implementation of the regional congestion pricing policy has greatly restricted the use of cars, and therefore the utilization rate of bicycles in the city of London is relatively high.

Inner London is a high-density residential area, employment, and service center, but due to the uneven distribution of public transport, bicycle use shows a correspondingly disproportionate distribution. The use of bicycles is relatively higher in less developed areas, with promotion of bicycle use by the government including developing bicycle roads and excellent lease management systems.

In the London suburbs, due to the low residential density, longer trip distances, except for a few areas, bicycle usage is lower than the City of London and inner London. To vigorously promote bicycle use, the London government has implemented many policies, including cooperative construction, developing and expanding the bicycle culture, construction of bicycle traffic infrastructures, enhancement of the security of cyclists, etc. The most representative three measures are described next.

9.6.1.1 Setting bicycle parking at bus stops

Since 2007, London has pushed ahead with the construction of bicycle parking around public transport stations such as subways and light rail, as shown in Fig. 9.22, encouraging citizens to use bicycles to get to the bus stop and then transfer to public transportation. For example, in July 2008, the first bicycle parking plot was built at DLR Shadwell station in Tower Hamlets.

FIGURE 9.22 Bicycle parking plot adjacent to a bus stop.

According to TFL's five criteria for convenient bicycle parking at DLR stations, Shadwell station has carried out the following specific construction measures [5]:

1. Dividing the bike parking areas;
2. Providing visible, durable, and transparent protection for parking areas;
3. Providing standard identification and information services;
4. Providing strong, safe, and convenient bicycle support for securing bicycles;
5. Setting up clear video monitoring near the entrance to the station.

After the bike park was built, bike usage in Shadwell increased by 50%. The successful experience of setting up bicycle parking lots in Shadwell station promoted the construction of bicycle plots along each DLR station and near bus stops, to increase the use of bicycles.

9.6.1.2 Implementation of the bicycle rental scheme

In 2010, London developed a bicycle rental scheme in the central area of the City of London, offering a bicycle rental service for travelers to London and surrounding areas, with users paying a small fee for a 24-hour bike rental service. TFL works with municipal boroughs, such as Camden, City of London, City of Westminster, Hackney, Islington, Lambeth, Kensington and Chelsea, Southwark, and Tower Hamlets, along with the royal parks and landowners to build bicycle docking stations. There is a rental point every 300 m, which improves the service level of the bicycle system [5]. The main objectives and contents of the bicycle rental plan are as follows:

1. Providing a sustainable, low-emission mode of transport;
2. Providing services for local residents' business travel and recreation, and for tourists;
3. Encouraging the use of bicycles for short trips inside the City of London;
4. Convenient bicycle usage, with a service available 24 hours a day and 365 days a year;
5. Easing congestion on subways and buses;
6. Providing an innovative direction for London's transport network;
7. Promoting the use of bicycles by car users;
8. Developing bicycle travel into a mainstream travel mode.

The bicycle rental scheme in the central area of the City of London began in July 2010 and took a step-by-step approach, available only to members for the first few months, and extended to other users after December 2010, allowing anyone to rent the bicycle. The number of bicycles increased steadily, and the covered area was expanded gradually. By March 2012, the bicycle rental program had been extended to the eastern part of the city, more than 8000 bicycles having been put into operation, resulting in a bicycle rental system consisting of more than 570 docking stations and 15,000 docking points. Fig. 9.23 shows bicycle docking points.

Fig. 9.24 shows the monthly rental quantity of the London bicycle rental plan since the time of implementation to March 2012 [8]. By the end of December 2011, there was an average daily rental quantity of 20,700, and the monthly rental quantity for December 2011 was 76 million.

FIGURE 9.23   Bicycle rental plot in London.

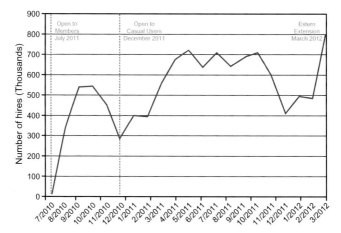

FIGURE 9.24   Number of rental bicycles in central London.

9.6.1.3   Construction of the cycle superhighway

To improve the traffic environmental conditions for cyclists, extend the bike trip distance, and enhance the security of bicycle travel, TFL worked with municipalities and enterprises, and built a cycle superhighway connecting residential areas of outer London with the center of London. In 2010, two of London's cycle superhighways "from Morton (Morton) to the city of London" and "from Barking to Tower Hamlets" were launched (No. 3 and No. 7 red lines in Fig. 9.25). A cycle superhighways network with 12 lines have been planned to be built (blue lines in Fig. 9.25).

As shown in Fig. 9.26, the cycle superhighways are provided with clear traffic signs on the road surface, free from interference of motor vehicles and pedestrians, which greatly improves the cycling safety and traffic environment, and attracts more people to use bicycles. The construction of cycle superhighways is of great significance in promoting the use of bicycles.

9.7   Digital UK

In June 2009, the United Kingdom released the Digital UK [18] plan, which clearly proposed transforming the United Kingdom into the digital capital of the world and consolidating the United Kingdom's dominant position in the digital knowledge-based economy. The main development goals included:

1. Making Digital Britain benefit all;
2. Improving and modernizing the national communications infrastructure to enable the United Kingdom remained invincible in the global digital economy;
3. Providing digital broadcasting platforms for broadcasters and listeners;

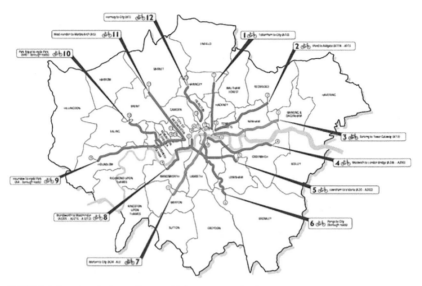

FIGURE 9.25   London's cycle superhighway network. *From Keegan M. Mayor's transport strategy. London: Transport for London; 2009.*

FIGURE 9.26   London's cycle superhighway.

4. Putting Britain at the center of the world's creativity;
5. Ensuring the quality and coverage of public services, obtaining information from multiple suppliers through multiple platforms to reach the world's top level;
6. Creating an environment for the cultivation of digital talents to inspire the next generation to become leaders in the field of creative research and technological innovation in the United Kingdom;

7. Ensuring that every British citizen lives and works online in a safe and secure manner and that their privacy is not violated;
8. Ensuring that public services provided in the United Kingdom are in line with users' expectations for new technology.

Examples of digital Britain include the following.

9.7.1 Urban intelligent transportation system

London is the first city in the world to adopt an intelligent transportation system. The railway transportation uses a global positioning system, and the traffic control center includes the location of all cars. Passengers on train platforms can always know the arrival time and terminal for the next train from a display board; sensors on the platform provide the number of waiting passengers to the control center, allowing dispatchers to control the number of trains and the time intervals between departure flexibly.

The London Underground network is divided into one to six zones from the city center to the outer edge. Passengers can buy single-zone or multi-zone tickets. To ease traffic pressure, tickets to or through the city center and during rush hours are more expensive. The London municipal government has built a uniform parking website, providing information such as the location, charge, and flow of open, underground, and internal parking lots in the city center. Car owners can reserve a parking space and time on the website before arrival and can make the payment by credit card.

9.7.2 The cloud city

Britain has the lowest Internet cost during the off-peak period in the world, and the Internet cost during peak hours is lower than the average of other OECD countries. London subway stations provide a wireless network. On the one hand, the provision of network services can improve service levels of the subway, on the other hand, it can help handle traffic information and ensure optimal traffic control.

9.7.3 Pilot applications of intelligent buildings in Gloucester

In 2007, a smart house was established in Gloucester, and sensors installed around the house. Information from the sensor allowed a central computer to control a variety of household devices. This smart house is equipped with a computer terminal as the core of the monitoring and communication network, using infrared and an induction cushion to automatically monitor the elderly walking around the house. A room was equipped with medical equipment for measurement of heart rate and blood pressure for the elderly residents, and automatically transmitted the results to the relevant doctors.

In addition, this smart house can provide air conditioning to maintain a set temperature and other functions.

9.8 Summary of the London and urban transport experience

1. Creating the urban planning masterplan, formulating the transport development strategy, and releasing the traffic operation monitoring report

London has published a series of reports, such as *The London Plan*, *Mayor's Transport Strategy*, and *Travel in London*, which have elaborated the government planning for urban transportation from different aspects. *The London Plan* makes a detailed analysis and planning of Greater London, including population size, economic development, environmental change, transportation system, urban space, etc. *Mayor's Transport Strategy* discussed the traffic strategy of the London mayor, including the vision, challenges, policy, strategy, and implementation. *Travel in London* analyzed the recent history of the traffic state, including traffic flow, mode sharing, spatial and temporal distribution, traffic network, traffic safety, environmental impact, bicycle use, logistics, and transportation. These reports guide the city and development direction of urban traffic, control the running of the city, and release timely information to the public.

Key points of the experience include:
a Formulating detailed urban planning plans to guide urban construction;
b Formulating a long-term transport development strategy to guide the development of transport;
c Conducting rich traffic operation monitoring, reflecting the traffic status;
d The government issued timely announcements and listened to the opinions of the public.

2. Establishment of a perfect public transport system and comprehensive transport system

London has a well-developed public transport system, covering the central area and suburbs of London, with a multimode operation above and underground. London's public transport system runs for a large part of the day, and according to demand the number of runs increases during peak hours. The different methods and routes of London public transport are connected by an integrated transport hub, which reduces the transfer costs of citizens, promotes the development of the integrated transport system, and makes a great contribution to the commuting and leisure of London citizens.

Key points of the experience include:
a High coverage of the public transport system, which can provide passenger services to the public in various areas;

b An excellent system of public transfer hubs, which can improve the service level of the public transport system;

c An intelligent public transport management system, which ensures safe, reliable, and efficient operation.

3. Vigorous development of walking and cycling systems

London's walking and cycling systems have begun to take shape, with many lines being built or planned. Walking and cycling systems serve as effective links between public transport systems and travel destinations, making outstanding contributions to the passage of citizens. They can also become an important part of citizens' leisure and entertainment, adding variety to the lives of citizens.

Key points of the experience include:

a Vigorously promote the use of walking and cycling systems to citizens; and

b Supporting facilities will be built to ensure a high level of service for walking and cycling systems.

4. Establishment of intelligent transportation systems

London's intelligent transport system serves different modes of transport. Through intelligent detection, detailed traffic information is obtained. Real-time and timely information is released through data analysis and processing, which can guide traffic management and the passage of users.

FIGURE 9.27 Ecology and harmony in St. James's Park, London.

FIGURE 9.28   An elaborate landscape in St. James's Park, London.

FIGURE 9.29   A duck house in St. James's Park, London.

Key points of the experience:
a Extensive information collection to control the urban traffic operation status and traffic demand distribution;
b Timely release of information, to provide information services for traffic management and travelers.

FIGURE 9.30   Perfect integration of building and plants.

The travel system development of London considers the landscape, environment and architecture, urban structure design, transportation system, ecology, and livability comprehensively. The environment is elegant and unforgettable, as shown in Fig. 9.27−9.30.

References

[1]   Baidu Baike. London, <http://baike.baidu.com>.

[2]   Wikipedia. London, <http://zh.wikipedia.org/wiki/London>.

[3]   Greater London Authority. The London plan: Spatial development strategy for Greater London. London: Greater London Authority; 2011.

[4]   Greater London Authority. 2011 Census first results: London boroughs. London: Greater London Authority; 2012.

[5]   Keegan M. Mayor's transport strategy. London: Transport for London; 2009.

[6]   Greater London Authority. Rising to the challenge: the Mayor's economic development strategy for Greater London, public consultation draft. London: Greater London Authority; 2009.

[7]   Department for Transport. Delivering a sustainable transport system. London: Department for Transport; 2008.

[8]   Transport for London. Travel in London Report 5. London: Transport for London; 2012.

[9]   Transport for London. Modes of transport, <http://www.tfl.gov.uk>.

[10]   Transport for London. Standard Tube map, <http://www.tfl.gov.uk>.

[11]   Transport for London. Transport for London Annual Report and Statement of Accounts 2012/13. London: Transport for London; 2013.

[12]   Transport for London. Docklands Light Railway System map, <http://www.tfl.gov.uk>.

[13] Transport for London. Docklands Light Railway mainline rail connections map, <http://www.tfl.gov.uk>.

[14] Transport for London. London Overground network map, <http://www.tfl.gov.uk>.

[15] Transport for London. London Rail and Tube services map, <http://www.tfl.gov.uk>.

[16] Transport for London. Tramlink user guide, <http://www.tfl.gov.uk>.

[17] Transport for London. Improving walkability. London: Transport for London; 2005.

[18] Department for Culture, Media and Sport, Department for Business, Innovation and Skills. Digital Britain Final Report. London: Office of Public Sector Information; 2009.

Further reading

Greater London Authority. Environment and Scientific Services, British Waterways. London: Greater London Authority; 2011.

Chapter 10

Amsterdam, The Netherlands

Chapter Outline

10.1 Overview of the city

Amsterdam is located in the northwestern part of the Netherlands, on the southwestern shore of Ijsselmeer Lake. It is the capital of the Netherlands, the largest city, the second largest port, and the largest industrial and economic center in the Netherlands. As of August 2013, the population of Amsterdam was 802,938 [1], with the population growth rate continuing to decline for three consecutive years. The total area is 219.32 km^2, of which the land area is 164.89 km^2 [2], and the population density was 4870 person/km^2. Amsterdam consists of seven districts, as shown in Fig. 10.1. The Centre District is the most densely populated and flourishing area, with a total area of 8.04 km^2, a land area of 6.28 km^2, and a population of approximately 81,000. The density is 12,900 person/km^2 [3], which is much higher than the overall average density of the city.

The Amsterdam Metropolitan Area consists of 36 cities including Amsterdam and its surrounding cities, including Almere, Amstelveen, and Haarlemmermeer. In 2012, the population was 2,330,000, with a total area of $2,580.28 \text{ km}^2$ [3] and a population density of 903 people per km^2. Table 10.1 shows a comparison of the population, land area, and population densities of the Amsterdam Metropolitan Area, Amsterdam, and its central district.

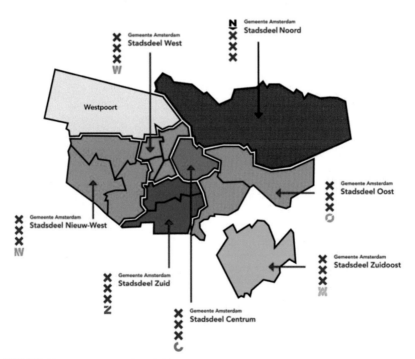

FIGURE 10.1 Amsterdam city division. *From http://www.iamsterdam.com/en-GB/living/city-of-amsterdam/amsterdam-city-districts.*

TABLE 10.1 Population, land area, and population density of Amsterdam, Amsterdam Metropolitan Area, and Amsterdam Central District.

	Amsterdam Metropolitan Area	Amsterdam City	Amsterdam Central District
Land area (km^2)	2580.28	164.89	6.28
Population (thousands)	2330	803	81
Population density (person/km^2)	903	4870	12,900

10.2   Urban structure and land use

Amsterdam is a beautiful city with a low and flat terrain, developed water system, and river channel. The water surface area covers about one-fourth of the area, with parks and green areas covering about 12% [4]. The urban architecture and human landscape are blended with the natural ecology of the city. The city center has a 17th century fan-shaped canal system. The Amstel river passes through the city center, spreads into the fan-shaped canals, and finally flows into IJ Bay, leaving a unique and beautiful fan-shaped water system to the city of Amsterdam, as shown in Fig. 10.2. Using its unique canal-shaped structure, Amsterdam has constructed beautiful buildings on both sides of the canal to form a fan-shaped urban structure. Many Renaissance Baroque buildings are included in the most fascinating landscape of Amsterdam, with the Amsterdam Canal District being listed on the World Heritage List.

The fan-shaped canals divide the central city into more than 90 zones. More than 1000 bridges connect these zones. Amsterdam can be described as a floating city. In the past, these canals were used for transport. Now, although these canals have lost their transport functions, they have remained as an important part of the historical heritage. As shown in Figs. 10.3 and 10.4, today's canals are representative of Amsterdam and are great places for citizens and tourists to enjoy and rest.

10.3   Public transport

Amsterdam has a developed public transport system with a clear hierarchy of services. The main public transport modes are subways, trams, buses, and water transport (Table 10.2).

The public transportation modes in the central city are mainly trams and buses. The trams in Amsterdam began operating in 1875 and there are currently 16 tram lines. The tram lines form a network with a total length of

FIGURE 10.2   Amsterdam urban land and water distribution map.

FIGURE 10.3   Canals in Amsterdam.

FIGURE 10.4   Canal landscape in suburban areas of Amsterdam.

TABLE 10.2 Travel mode share in Amsterdam city.

Mode	Bicycle	Public transport	Private car	Green traffic total
Share (%)	34	26	40	60

80.5 km [5], and there are 237 trams. The tram vehicles are long, as shown in Fig. 10.5, at about twice as long as a regular bus and are therefore able to carry more passengers. As shown in Fig. 10.6, the trams provide a good environment and maintain comfortable service quality even on peak hours. There are 30 bus routes in Amsterdam. Bus stations are spread more densely. The bus system has a high punctuality rate and is an effective supplement to the trams.

There are 4 subway lines in Amsterdam with an operating network of 42.5 km [6], with 52 stations and a site spacing of 0.82 km. As shown in Fig. 10.7, three of the four subway lines start from the central station and extend radially to the southeast and southwest. The other line is semiannular from northwest to southeast. The subway connects the central city of Amsterdam with surrounding suburbs and other cities in the metropolitan area. Transfer between subways and other public transport modes is very convenient.

FIGURE 10.5 A tram in the city of Amsterdam.

FIGURE 10.6 The pleasant traveling environment in tramcars.

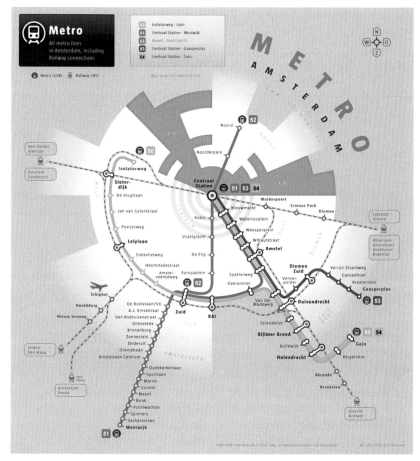

FIGURE 10.7 Amsterdam metro line map. *From http://transitmaps.tumblr.com/post/ 20463633551/amsterdam-circle.*

Amsterdam has developed water transport, with eight ferry routes in the city. Among these, there are five ferry routes across IJ Bay, providing more choices for citizens. In 2012, the average passenger volume of the eight ferry lines per day was 42,500 [7].

10.4   Construction of the bicycle transportation system

Amsterdam has one of the highest proportions of bicycle trips in the world. The bicycle share rate is as high as 34%, and the bicycle share rate in the central city reaches 60%. In 2012, there were about 880,000 bicycles [8] in Amsterdam, at 1.1 bicycles per capita. Bicycles are very popular. The

FIGURE 10.8 Bicycle lanes in Amsterdam.

popularity of bicycle travel in Amsterdam benefits from the city's comprehensive bicycle transport system, comfortable riding environment, and a deeply rooted bicycle culture.

1. Complete bicycle-specific roads

   Amsterdam has built 400 km of bicycle lanes [9], as shown in Figs. 10.8 and 10.9. Some bicycle roads are paved with colored asphalt, with clear signs and an excellent riding environment.

2. Set up large bicycle parking lots

   At the central station, there is a large bicycle parking lot, as shown in Figs. 10.10 and 10.11, which is convenient for cyclists transferring to public transportation. In contrast, parking fees for cars are very expensive, promoting people's willingness to choose bicycle + public transportation.

3. The intersection provides a dedicated waiting area for bicycles

   As shown in Fig. 10.12, the signed waiting area for bicycles is set at the intersection, and the stop line for cars is behind the bicycle waiting area, thus bicycles can stop in front of cars while waiting for traffic signals.

4. Convenient bicycle rental system

   Amsterdam has a wide range of rental bicycles for visitors' convenience. There are also small bicycles especially designed for children. As shown in Fig. 10.13, tourists ride on rental bicycles to explore downtown attractions.

FIGURE 10.9   Color paved bicycle lane in Amsterdam.

FIGURE 10.10   Bicycle parking lot near Amsterdam Central Station. *From http://en.wikipedia. org/wiki/File:Bikes_parking_in_Amsterdam_Central_Station_.JPG*

Amsterdam is not a car-prioritized city. When we look at the passage space, the parking space, the configuration of signals, and the setting of waiting areas, we find that bicycles have great priority throughout the city. It can

FIGURE 10.11 Bicycle parking adjacent to public buildings.

FIGURE 10.12 Dedicated signal waiting area for bicycles.

be said that bicycles are the "darling" of the city. People love bicycles, and riding bicycles has become a deliberate choice. In terms of natural conditions, Amsterdam is on an alluvial plain, and the flat terrain provides good conditions for widespread bicycle use.

FIGURE 10.13   Tourists use rental bicycles to visit the city.

Bicycles as a green method of transportation also bring many benefits to Amsterdam, not only reducing vehicle emissions and pollution, which contribute to environmental protection, but also improving traffic safety. In 2007, the number of deaths due to traffic accidents in Amsterdam was only 18 [10].

10.5   Case of eco-city construction: Almere

Almere is a small city in the Amsterdam Metropolitan Area. It is located 35 km east of central Amsterdam. Its development began in 1967 and it became a municipality in 1984. It is one of the newest cities in the Netherlands. Almere has a total area of 248.77 km^2, of which the land area is 129.6 km^2. As 30% of the city is green areas, it is a city with a very good natural ecological environment. However, the city's ground plane is 4 m below sea level, and so it was necessary to build a dam to prevent seawater intrusion.

Since its development in 1967, the city of Almere has attracted 195,000 residents and 14,500 commercial enterprises, making it the seventh largest city in the Netherlands. The population density of the central city is 3000 people per square kilometer. At the beginning of the construction of the new city, the residents were mainly of low- and middle-income groups, and most of were young. Two-thirds of the residents are between 20 and 65 years old. In recent years, the demographic has entiered a stage of transition.

Almere's development goal is to build an eco-city with energy conservation, environmental protection, and balanced employment and residence. Therefore from the beginning of its development, it has focused on attracting enterprises. At present, there are 14,500 companies, providing about 90,000 jobs. This is about two-thirds of the city's working population. At present, about 50% of the population living in Almere work in the main city of Amsterdam, and about 30% work in Almere. The balance of employment and residence is still being advanced.

Almere's biggest highlight in the planning and design of urban land is the three-levels hierarchical planning and construction in the central city, namely the underground traffic layer, the ground commercial layer, and the top residential layer.

1. *Underground traffic layer*: for motor vehicles and bicycle roads and parking lots. There are entrances to the underground traffic layer outside the central commercial area. Vehicles can enter the layer directly, as shown in Fig. 10.14. In the central commercial street, escalators connect the ground floor with the underground traffic layer, as shown in Fig. 10.15. At the same time, it can be seen that the above-ground buildings are directly connected by corridors. This three-level development mode is constructed to achieve complete separation of pedestrians and vehicles.

2. *Ground commercial layer*: the ground floor is used to build shopping streets and office buildings, which greatly enhances the popularity and overall vitality of the area. In order to create a relaxed and enjoyable shopping environment, rest places are set in the shopping street, as shown

FIGURE 10.14 Entrance to the underground traffic layer.

FIGURE 10.15   The ground floor of the central commercial street connects to the underground traffic layer by escalators.

FIGURE 10.16   Setting up a rest area to create a relaxed shopping environment.

in Fig. 10.16. Opened squares, fountains, and open-air cafes are built in the central area of the city to create a beautiful landscape, as shown in Figs. 10.17 and 10.18. Some shopping streets have a roof, which provides a good environment for pedestrians, as shown in Fig. 10.19.

FIGURE 10.17 A wide square in an office area.

FIGURE 10.18 Opened squares, fountains, and open-air cafes in the center.

3. *Top residential layer*: the upper floor of commercial stores or office buildings is used as a residential building. There are corridors connecting each building to create a three-dimensional passage, as shown in Figs. 10.20 and 10.21.

FIGURE 10.19   Using a roof above the shopping street to improve the shopping environment.

FIGURE 10.20   Residential buildings are built on the upper floors of commercial shops, and interconnected by air corridors.

Through the development and construction of the above three-level design, vehicles and pedestrians are completely separated, creating a safe and warm living, shopping, and working environment. At the same time, the

FIGURE 10.21 Residential buildings above commercial shops.

mixed development in the central area of Almere is promoted, which benefits job—housing balance and promotes commercial and residential integration. This is a lesson worth learning in the construction of ecological cities.

10.6 Amsterdam's urban design and transport development experience

The city of Amsterdam pays attention to planning and features—canals extend in all directions, the architecture is characterized, the bicycle culture is deeply rooted in people's minds, the walking and bicycle systems are comprehensive, especially in large-scale new development areas, such as Almere, which show many advanced planning and design concepts.

1. Bicycle culture is deeply rooted in people's minds
   In Amsterdam, cycling has become fashionable. From the mayor to ordinary citizens, they all love bicycles. There are even bicycles in wedding motorcades. In the period of rapid economic growth and motorization trend, citizens of Amsterdam have fought against some policies in order to maintain a good bicycle-friendly environment. Cycling has become a conscious choice in Amsterdam. The perfect bicycle access and convenient bicycle parking facilities provide good conditions to travel by bike in Amsterdam.
2. Attach great importance to architectural and urban design

No matter whether in the old town of Amsterdam or in the new towns being built, pursuit of architectural art and streetscapes by the people of Amsterdam is clearly evident. The design is exquisite, the concept is advanced, the design is fine, and the buildings and landscapes will retain their beauty for a long time. The city illustrates precious experiences of urban civilization.

3. The canal system has become a major feature and highlight of Amsterdam

Amsterdam's complete river system creates a unique landscape in the city. People can browse and enjoy the cityscape of Amsterdam through boats on the canal. Along the canal, the buildings are beautiful and distinctive, with developed business facilities and a convenient living environment.

References

[1]   Statistics Netherlands: CBS Amsterdam Bevolkingsontwikkeling; regio per maand. <http://statline.cbs.nl/StatWeb/publication/?DM = SLNL&PA = 37230ned&D1 = 0,17&D2 = 39,66,88,121&D3 = 91-95&HDR = T&STB = G2,G1&VW = T> [accessed 14.12.13].

[2]   Gemeente Amsterdam Bureau Onderzoek en Statistiek: Area, population density, dwelling density and average dwelling occupation, <http://www.os.amsterdam.nl/tabel>; January 1, 2012 [accessed 14.12.13].

[3]   Centraal Bureau voor Statistiek. <http://www.cbs.nl/nl-NL/menu/home/default.htm> [accessed 14.12.13].

[4]   Wikipedia. Amsterdam, <http://en.wikipedia.org/wiki/Amsterdam#cite_note-12percent-36> [accessed 14.12.13].

[5]   GVB. Transport in figures, <http://www.gvb.nl/english/aboutgvb/facts-and-figures/Pages/transport.aspx> [accessed 14.12.13].

[6]   Wikipedia. Amsterdam Metro, <http://en.wikipedia.org/wiki/Amsterdam_Metro> [accessed 14.12.13].

[7]   GVB. Jaarverslag GVB Holding NV 2012. Amsterdam; 2012.

[8]   City of Amsterdam. Understanding Amsterdam. Amsterdam; 2013.

[9]   Cycling in Amsterdam. <http://www.amsterdamtips.com/tips/cycling-in-amsterdam.php> [accessed 14.12.13].

[10] Research and Statistics Division. Core numbers in graphics: fewer traffic deaths. Safety and Nuissance (in Dutch). City of Amsterdam. Archived from Wikipedia Cycling in Amsterdam. <http://en.wikipedia.org/wiki/Cycling_in_Amsterdam> [accessed 14.12.13].

Chapter 11

Munich, Germany

Chapter Outline

11.1 Overview of the city

Munich is the capital of Bavaria, and is the third largest city in Germany after Berlin and Hamburg. It is also one of the main economic, cultural, technological, and transportation centers in Germany and one of the most prosperous cities in Europe [1], housing the headquarters of Siemens, BMW, and MBB.

Eco-Cities and Green Transport. DOI: https://doi.org/10.1016/B978-0-12-821516-6.00011-4

TABLE 11.1 Overview of Munich's population and area in 2012.

	Area of Munich	Metropolitan area of Munich
Area (km^2)	412	311
Population (10,000)	148.5	139.5
Population density (per km^2)	3604	4486

Munich carried out an administrative division reform in 1992. At present, Munich is divided into 25 districts, including Altstadt-Lehel, Ludwigs vorstadt-Isarvorstadt, with a total area of 311 km^2 [2]. In 2012, the total population of Munich was about 1,395,000, with a population density of 4486 people per square kilometer. The population density data of Munich region and urban area are shown in Table 11.1.

11.2 Urban structure and land use

Munich has gradually developed from the old town. It is a typical city which has changed from a concentric circle expansion mode to a long-axis development mode. The urban spatial structure has formed a "point-axis-network" system [3]. Munich consists of three parts: the metropolitan area, inner city, and outer city.

Munich is a representative city which combines transportation planning and land use organically. It has abandoned the combination of radioactive highways and urban routes. For many years, it has devoted itself to changing people's way of traveling through strategic land use planning. Munich's new large-scale urban development projects are mostly implemented around bus stations, while other small-scale development projects are maket-driven [4].

The urban development of Munich mainly focuses on the periphery of public transport corridors, that is, Munich compactly develops with the public traffic stations and line as its center. The compact and complex land use model greatly reduces people's travel distance and is a reasonable method of organizing travel. At the same time, a good public transport system also reduces the use of urban road land, greatly improving the urban environment. Thanks to its developed public transport system, Munich became a famous "green city" in Europe. Munich's urban development principles are "compact, urbanization, and green," which are embodied in the following:

1. Compact: transfer from urban expansion to urban renewal. Higher density land redevelopment in the built area can save limited land resources, reduce the use of cars, and ensure the effective operation of public transport.

2. Urbanization: land use with complex functions, mixed with living, work, and entertainment. On the one hand, this can reduce people's travel distances, and on the other hand, it can help create a diversified atmosphere for urban life.
3. Green: urban sprawl has been controlled effectively, so farmland and green land on the edge of the city can be protected. Meanwhile, a wide range of public green space and tree planting within the city (Munich has the largest urban park in Europe) has improved the quality of urban life [4].

The compact urban development model guided by public transport not only frees the city from the plight of traffic congestion, restores the quiet and clean of the city, but also creates a compact urban space suitable for walking and communication. Munich is recognized as a representative of a sustainable development city throughout the world.

At the same time, in the process of urban space expansion, Munich attaches great importance to the historical features of the old city, stressing the nature and continuity of urban development.

11.3 Mobility and traffic demand characteristics

Munich has 530 cars per 1000 people. As Munich has an efficient, fast, and well-served public transport system, people choose public transport as the main mode of travel. In Munich, the share of green traffic is 63% and that of private cars is 37% (see Table 11.2). Munich is a city dominated by green transportation.

11.4 Development of public transit

Munich has a mature urban public transport system consisting of the U-Bahn, S-Bahn, Trambahn, and bus. In the construction of its public transport system, Munich has adopted a variety of public transport modes, clear line system, humanized site design, and line network layout coordinated with urban form.

The layout of the metro and urban light rail is set according to the local population distribution, and the two levels of lines superimposed coincide with the population density distribution. The metro and light rail are linked

TABLE 11.2 Traffic mode share in Munich (2008).

Mode	Walk	Bicycle	MRT + LRT + commuter train + bus	Private car	Total of green transportation
Downtown of Munich	28%	14%	21%	37%	63%
Munich city	25%	13%	15%	47%	53%

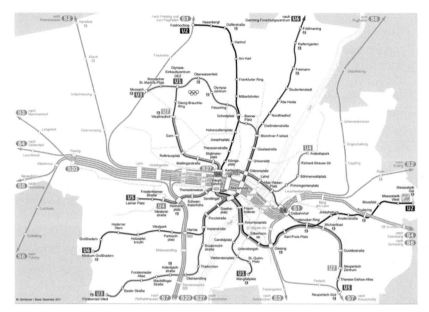

FIGURE 11.1 Network map of Munich metro (U-Bahn, 2013). *From http://zh.wikipedia.org/wiki/File:U-Bahn-Plan_M%C3%BCnchen.png.*

together. When the train enters the city, it enters the underground. When it leaves the city, it runs over ground. Because the radiation rays are interwoven in the downtown area, each line intersects with other lines more than twice, so that it is very convenient for people to transfer between lines.

11.4.1 Munich metro

Munich's U-Bahn system currently consists of seven routes, 100 stations, and has a total length of 103 km [5]. The service radius of the metro network is about 10 km. Metro lines basically cover the entire city and some suburbs of Munich, and it is the core framework of the urban public transport. The number of passengers transported per year is 3.78 million, with the speed of the trains reaching 80 km/h. The Munich metro network is shown in Fig. 11.1.

Each subway station is marked with a white "U" sign on a blue background. The nearest Munich metro station can be found by looking for blue and white U-shaped signs.

During the day, the metro departs according to accurate passenger flow statistics. The departure time is every 3−5 minutes in peak periods and every 10 minutes at other times. The departure time is every 20 minutes on Sundays and other holidays. There is no service from 2 to 4 a.m. The departure time is not more than 40 minutes [3].

FIGURE 11.2   Munich metro (U-Bahn, 2013).

Interstation distance in the Munich Metropolitan Area is about 900 m. Stations in metropolitan areas are denser, with an interstation distance of less than 500 m; and the interstation distance in pedestrian areas around Marion Square is less than 350 m. The short interstation distance stimulates Munich citizens' enthusiasm to use public transport [3].

Each metro station in Munich has its own unique design style, with various colors and decorations, which have strong identifiability. Metro stations built in recent years are especially unique, with distinctive use of color and illumination. Some stations have light purple backgrounds, some orange-red backgrounds, and others light green backgrounds. In terms of decoration, some look abstract and some are modeled with industrial structures. Some feel warm and intuitive, some feel cold, lines are prominent, and some platforms are designed in a traditional style. In short, each subway station has its own color and style. Passengers can even remember stations without remembering its name, by recalling their particular color and decorative style (Fig. 11.2).

11.4.2   Munich S-Bahn

There are seven light rail lines (S-Bahn) in Munich, which are operated by DB Company. This is the backbone network of the bus system in Munich.

S-Bahn connects Munich's inner city with its surrounding suburbs and towns. Its maximum service radius is about 40 km, and it carries an average of 800,000 passengers per day, traveling 20 million kilometers per year [3].

The departure frequency of S-Bahn is usually every 20 minutes, and 40 minutes during the night and at weekends [3]. The distance between Munich light rail stations varies from 1.5 to 7.5 km depending on the distribution of suburban settlements. Light rail stations generally include bus stops and parking lots (including cars and bicycles), and are not far from community centers or small town centers. Munich has become Europe's favorite city for public transport [3].

11.4.3   Munich tram

Trams are used to serve urban areas with medium population density, and areas where the subway has not yet been built, and is an important supplement to the metro. As shown in Fig. 11.3, the Munich tram system has 10 lines with a total length of 71 km.

The distance between tram stations is usually about two blocks (about 400 m) [3]. The frequency of tram departure is close to that of the metro, which is 5 minutes in peak time, and 10 minutes the rest of the time (Figs. 11.4−11.6).

11.4.4   Buses and taxis

Munich has 65 bus lines [6], totaling 458 km in length and carries more than 1 million passengers a day and more than 4.8 billion a year (shown in Fig. 11.7).

Munich's public station density is high. This ensures that there is a bus stop within 400 m. Residents only need to walk for a few minutes to reach a subway station, tram station, or bus station.

Trams and bus stops are marked with a green "H" on a yellow background.

Munich's public transport system is jointly managed by the Munich Transport System (MVV) and the German Railway (DB). The first bus runs from about 5 a.m. and the last bus varies between 6 p.m. and 1 a.m. the next day.

The expansion of public transport has been accompanied by the continuous improvement of the supporting service system. In Munich, bus routes, bus stops, timetables, and ticket vendors are well integrated, greatly reducing the average passenger transfer time. In mixed traffic, public transport is given priority by traffic signals and traffic regulations. Many stations have been renovated and have a barrier-free design. Disabled people can easily enjoy convenient public transport services without the assistance of others (Table 11.3).

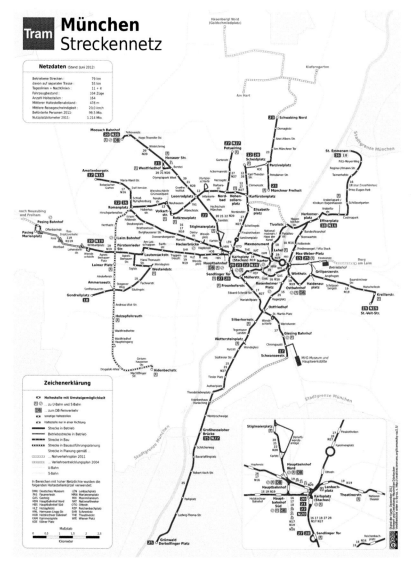

FIGURE 11.3   Network map of the Munich tram system (2012). *From http://commons.wikimedia.org/wiki/File:Stra%C3%9Fenbahnnetzplan_M%C3%BCnchen.png.*

Munich is a bicycle-friendly city, and the Munich government actively encourages people to ride bicycles. Since 1980, Munich's bicycle network system has doubled; it now has 800,000 bicycles and 1000 km of bicycle lanes. The average Munich family has at least two bicycles.

Munich's streets are generally narrow, especially in the old town areas, but all have bicycle lanes, mainly using color to identify them, while busy

FIGURE 11.4 A Munich tram. *From http://www.douban.com/photos/photo/1859140658/large.*

FIGURE 11.5 A tram running in Munich.

commercial areas, subway stations, and railway stations also have set bicycle parking areas. Figs. 11.8 and 11.9, respectively, shows a bicycle parking area near a metro station and one at the central railway station.

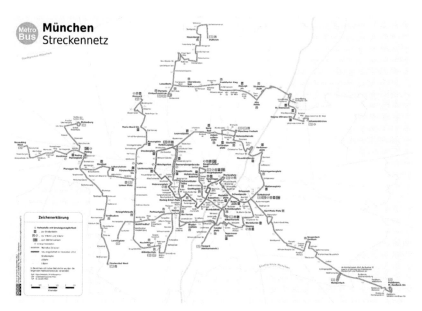

FIGURE 11.6   A convenient tram at the central railway station.

FIGURE 11.7   The bus network in Munich (2012). *From http://commons.wikimedia.org/wiki/ File:Metrobusnetzplan_M%C3%BCnchen.png.*

TABLE 11.3 Technology data for railway transportation in Munich.

Name	Length (km)	Annual passenger traffic volume (billion)	Load intensity (persons/ day · km)	Number of stations	Average interstation distance
MRT (U-Bahn)	95	0.378	10,901	140	0.96
S-Bahn	442	0.78	4835	148	2.99
LRT	79	0.104	3606	165	0.48

FIGURE 11.8   Bicycle parking area near a subway station.

Bicycle riding has become a fashion in Munich. During rush hours, a huge bicycle flow is part of the street scene. German subways and trains have bicycle compartments, bicycles can be taken on the subway, and bicycle parking areas are set up at various public transport stations to facilitate the transfer of bicycle users. Adult and young riders can be seen at any time in the streets and alleys, as shown in Fig. 11.10. The shapes of the bicycles are also rich and colorful, as shown in Fig. 11.11, which shows a multipassenger bicycle. People use bicycles for leisure and recreation on holidays.

FIGURE 11.9   Bicycle parking area in front of the central railway station.

FIGURE 11.10   Adults and children enjoy bicycles.

Munich has established a perfect system named Call-A-BIKE for bicycle rental. Bicycles can be rented by dialing numbers on mobile phones or registering at the callabike.de website. The system can automatically display the

FIGURE 11.11   A comfortable group of cyclists.

locations of bicycles available, and guide users to the nearest bicycle efficiently. After arriving at the destination, bicycle can be left.

It has been reported that the Munich government plans to make the city the most bike-friendly city in Germany. Cycling is an excellent form of exercise, it can save energy, and can also promote environmental protection, forming a healthy and vibrant urban society. According to the estimations of Quande Bicycle Club, by vigorously promoting bicycle traffic, emission reductions will be at least 3 million tonnes per year.

Munich is also a very walkable city. Since 1996, Munich has adopted a design principle of building a pedestrian-friendly city. It is hoped that people will choose walking more when distance permits. These measures include paying more attention to pedestrians when designing streets and squares, improving pedestrian crossings on major roads, and providing more comfortable walkways. In order to protect the historic center of the city, maintain the quality of life of residents, and attract tourists, Munich has established a fully pedestrianized center. Many other streets and squares have strict prohibitions on cars [4]. Fig. 11.12 shows a comfortable pedestrian road and Fig. 11.13 shows a pedestrian mall.

In addition to the above measures, Munich is also restricting parking in order to encourage people to use public transport. Almost all free parking places in the city have been removed. The parking fee in the central business district is about 25 euros per day, which is about eight times the price of a 1-day public transport ticket (allowing the user to use the subway, tram, and bus at will). At the same time, high oil taxes have encouraged many Munich citizens to give up their cars and use public transportation.

FIGURE 11.12   Comfortable walkways in Munich.

FIGURE 11.13   Commercial pedestrian streets in central Munich.

To really solve the traffic problem, the fundamental strategy is to change people's travel ideas. Munich arouses public awareness of environmental protection through public education. The benefits of public transportation, bicycles, and walking are widely advertised on large public posters and electronic displays: more environmentally friendly, healthier, and safer.

11.5 Summary of the characteristics and traffic experience

1. Mixed land use reduces travel distance. Through mixed land use, residential buildings are close to jobs, and living facilities in residential areas are perfect. People can go to work and shop nearby. This compact and bus-dominated land development model has a very important reference significance for the land use model of cities in China.
2. Fully respecting nature and natural integration, attention is paid to hydrophilic and ecological design. As shown in Figs. 11.14 and 11.15, the English Garden in downtown Munich has an elegant environment and natural ecology, where people can surf and swim freely, bask in the grass, relax and play, enjoy delicious food, and chat with friends in elegant restaurants. The whole city can feel the emphasis on ecology, landscape, and beautification in design everywhere.
3. Munich is a paradise of green transportation. There is a wide variety of public transport systems, which can be seen everywhere. Whether it is walking areas, urban attractions, or public facilities and space, public transport can be accessed everywhere, enhancing the beauty of Munich.

FIGURE 11.14 People relax on the banks of Isar river in the English Garden.

FIGURE 11.15   A water-loving environment on the banks of Isar River in the English Garden.

FIGURE 11.16   Leisure space on the streets of Munich.

Safe, continuous, warm walking and bicycle spaces ensure that people are able to enjoy the unlimited comfort of using green traffic (Figs. 11.16−11.20).

FIGURE 11.17 Beautiful and serene waterfalls on Isar River in the English Garden.

FIGURE 11.18 Young people surf on Isar River in the English Garden.

4. Advanced traffic concept and traffic civilization. The citizens of Munich love bicycles, walks, and buses because of their love for nature, their pursuit of blue sky and white clouds, and their emphasis on environmental quality and health. The general mood of the whole society and a deep

FIGURE 11.19 Beautiful surroundings on the second floor of a beer house in downtown Munich.

FIGURE 11.20 Characteristic building in downtown Munich: BMW headquarters building.

understanding and pursuit of green transportation by decision makers fills the entire city with ecological and green passion and action, which is a fundamental solution to urban traffic problems.

References

[1] <http://baike.baidu.com/link?url = srOUmCN04BKE03fFm-C87KEXcD_gh4Ex-a4dMB_YZLUfiXvicxwIHwDfaKacpVRVy2lIdWarWP2rjOLjJgD8Na>.
[2] <http://www.baike.com/wiki/%E6%85%95%E5%B0%BC%E9%BB%91>.
[3] Yuanyuan GU. Study on the characteristics of Munich public transport system. Plan Forum 2012;(28).
[4] Zhang Z. Towards a compact urban form: Munich as a bus city. Annual Academic Conference on Urban Planning in China; 2004.
[5] Site Wang. Munich metro: convenient transportation. China National Daily; 2005 (004 edition).
[6] <http://you.ctrip.com/traffic-d572-munich.html>.
[7] Shaocai L. Low carbon travel in Munich. Sports Artic Sci Tech. 2010/11.

Further reading

<http://zh.wikipedia.org/wiki/%E6%85%95%E5%B0%BC%E9%BB%91>.
<http://newspaper.jfdaily.com/jfrb/html/2008-03/31/content_401252.htm>.

Chapter 12

Madrid, Spain

Chapter Outline

12.1 Overview of the city

Madrid (the capital of the Madrid Autonomous Region) is located in the central part of Spain and is at the geometric center of the Iberian Peninsula in western Europe. Madrid is 667 m above sea level, and is the highest capital city in Europe. In the 16th century, Madrid was still a peaceful agricultural community in the Sierrra De Guadarrama, far from the sea, with a hot climate and dry air. King Philip II moved the palace from Toledo to Madrid in 1561, and from that time Madrid's urban status was upgraded. As the capital, the pace of construction was accelerated. Today Madrid, as the largest city in Spain, has become the political, economic, cultural, and transportation center of the country. In 2010, Madrid's population reached 3.27 million, covering an area of 606 km^2, with a population density per square kilometer of about 5400 [1].

The word "Madrid" has two representative meanings. The Madrid Autonomous Region (see Fig. 12.1) is a single province and autonomous region of Spain whose capital is the city of Madrid. Outside Madrid, the inner and outer rings are divided according to the distance from Madrid [2]. The population and area of each region are detailed in Table 12.1.

Eco-Cities and Green Transport. DOI: https://doi.org/10.1016/B978-0-12-821516-6.00012-6

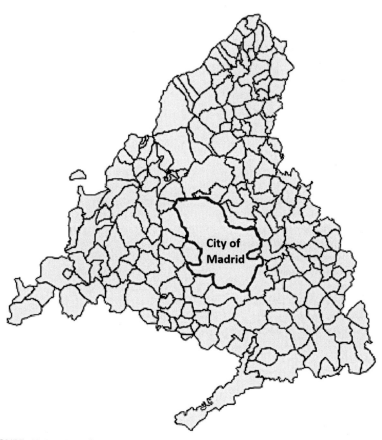

FIGURE 12.1   Map of Madrid Autonomous Region. *From http://en.wikipedia.org/wiki/File: Distritos_de_Madrid.svg.*

TABLE 12.1 Area and population of Madrid [3].

	Resident population				Urban area (km²)	Population density/ km²
	1986	1996	2001	2010	2010	2010
City	3,058,182	2,866,850	2,938,723	3,273,049	606	5398
Interior loop	1,537,472	1,913,804	2,182,688	2,370,337	2280	1040
Outer loop	184,918	241,636	301,973	815,298	5141	16
Autonomous Region	4,780,572	5,022,288	5,423,384	6,458,684	8026	805

TABLE 12.2 Mode split of residents in Madrid [4].

Mode	Walking	Public transport	Private cars	Green transport
Mode split (%)	26	35	36	61

12.2 Urban structure, mobility development, and traffic demand characteristics

From the outline of Madrid city and its road network, we can see that Madrid presents a city radiating from the central main city to the surrounding suburbs. In the 1980s and 1990s, industrial zones were built in the suburbs. Hence the city was expanded to the outer areas. However, at that time, the suburbs were only relatively simple industrial land with limited living facilities, and the residents remained concentrated in the city center. In recent years, with the development of rail transit in Madrid, residents have gradually moved to the suburbs, and the population gap between the suburbs and the main urban areas is narrowing (see Table 12.1).

In Madrid, with the expansion of population and land use, private car ownership continues to grow. Meanwhile, public transport has attracted residents effectively by large-scale construction and continuous improvement of facilities. The data in Table 12.2 shows the modal share of Madrid residents. The total ratio of green traffic (walking and public transport) is 61%. According to these statistics, the main mode of travel in the central urban area is walking; in the suburbs, public transport and private car travel account for about 50%; and in the far suburbs, private car travel is mostly used.

12.3 Development of public transit

The construction of Madrid's public transport system began in the 1940s and continued to intensify from 1970 to 1985. Under multiple modes of government-led and market-oriented management, a virtuous cycle of public transport system improvements has been established. A multilevel and efficient urban transport network has been formed, with a mature organizational structure and management system. The Madrid Transport Authority (CRTM) was established by the government in 1985, and its relationship with other agencies and operators is shown in Fig. 12.2.

In the Madrid Autonomous Region, the daily traffic volume is about 10 million people. Four major transport networks (suburban railways,

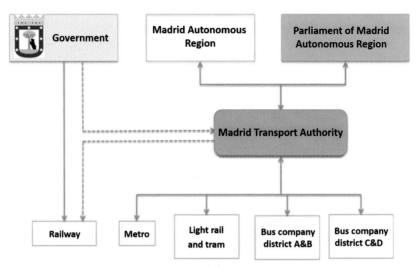

FIGURE 12.2 Traffic management system in Madrid [5].

subway, light rail and tramways, and buses) account for about 50% of the total traffic. The annual volume of the 310 km metro network is 700 million passengers. Every year, 200 million people use the suburban railway network and 450 million use the urban public transport network. The intercity bus network consists of more than 450 routes [5].

There are two modes of public transport within the interior loop: on-site buses and underground railways. Suburban train networks are provided for travelers between the interior loop and outer suburbs. Detailed descriptions of the different modes of public transport are presented next.

There are two kinds of ground buses: normal buses for citizens and sight-seeing buses for tourists. Madrid's bus network is highly developed, and the metropolitan area is almost completely covered. Clear route signs can be seen at every bus stop.

The Empresa Municipal de Transportes de Madrid (EMT) bus network system has been adopted in Madrid. The proportion of buses is quite large, second only to that of the metro in resident daily travel. By 2013, there were more than 2000 vehicles and 224 operation lines in the urban bus system. Each year, more than 400 million passengers travel by bus, with the driving distance reaching 100 million kilometers [6]. The buses network in Madrid is an important supplement to the metro. To meet the short-distance transfer needs of travelers, the distance from each bus stop to the nearest metro is under 100 m.

Official online support of the bus system is provided for the public's inquiries about bus operation information in Madrid. The official bus network can be found by links from the Madrid city official website (http://www.emtmadrid.es/Home.aspx), which includes bus routes, sites, latest developments, and pages that support self-inquiry and dynamic navigation (see Fig. 12.3).

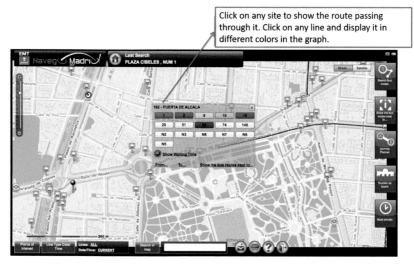

FIGURE 12.3   Bus navigation system. *From http://www.emtmadrid.es/Home.aspx.*

In order to meet the needs of bus accessibility, reconstruction of the M-30 loop was completed in 2006. This is the third loop in the city and one of the most important civil engineering projects in Europe in recent years. The M-30 loop is primarily the outer loop of Madrid. In order to meet the needs of different regions, different modes and capacities have been designed. In the process of reconstruction, the M-30 loop has successfully solved the problems of complex interchanges and chaotic lines in the past. Road green belts were built to add a touch of greenness to the city. Underground passages were built along the river area to avoid secondary damage to the riverbank after the overdevelopment of industry [5].

Fine loop construction and developed bus navigation have reduced the urban congestion in Madrid. In most cases, the traffic on the main urban road is relatively smooth and there is no serious congestion. The traffic status of Akala Street, the main road in Madrid, is shown in Fig. 12.4. Fig. 12.5 shows Madrid's highway leading to the northern suburbs, with normal motor lanes on both sides and priority roads for multipassenger cars in the middle.

Tourist buses are also an important part of Madrid's urban bus service. Tourists can not only use the bus to visit the Madrid Metropolitan Area, but also enjoy the sunshine in the open-topped vehicle, as shown in Fig. 12.6.

12.4   Underground railway

Madrid's underground railway system was unveiled by King Alfonso XIII on October 17, 1919. Until 2011, there were 281 stations with 27 two-line transfer stations, 12 three-line transfer stations, and 1 four-line transfer station

FIGURE 12.4   Akara street.

FIGURE 12.5   Highway from downtown to the suburbs.

(Avenida de America). The entire metro network consists of 12 main lines and 1 branch line, which are distinguished by different colors. The total length of the network is 281.78 km. Three stations offer interchanges within the platform, and 21 stations offer interchanges with the Madrid suburban

FIGURE 12.6   Tourist bus in Madrid.

FIGURE 12.7   Taxis queuing at fixed stations in downtown Madrid.

line. In addition, Madrid metro also includes three light rail lines (ML) with a total length of 27.78 km and 38 stations. There are 319 stations in the metro and light rail, and the total length is 309.544 km. According to this mileage, Madrid metro ranks among the top 10 in the world (Fig. 12.7).

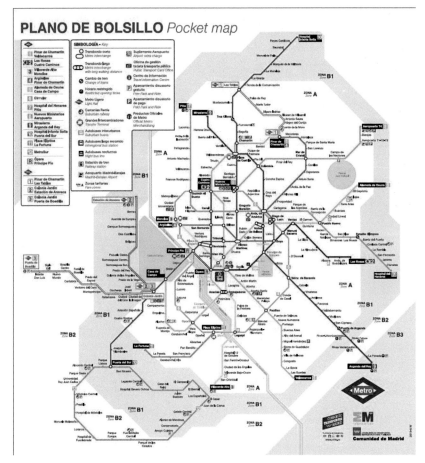

FIGURE 12.8 Madrid metro line map. *From http://www.metromadrid.es/.*

At present, the radiation range of Madrid metro network is quite large. One-third of its line mileage is beyond the urban area, which greatly facilitates travel between the urban area and the surrounding towns. In addition, the metro network connects the main external traffic nodes to facilitate passenger transfer, such as railway stations and Madrid International Airport. The layout of the metro lines is shown in Fig. 12.8.

The design of Madrid metro is quite distinctive. Because of the different construction ages, different architectural styles are clearly visible on the new and old metro lines. There are long (115 m) and short platforms (60–90 m). According to their shape, they can be divided into three types: island platforms, side platforms, and island-side platforms. Island-side platforms enable passengers to get off from different positions (e.g., from both sides and from the center), which can improve the turnover efficiency and speed of the flow of people. This transfer method has also been adopted in Xizhimen subway station in Beijing.

There are two types of transfer platform for the metro: the station for transfer between metro lines, and the hub for transfer to various other modes of transportation. Most passengers go through underground passages during transfer, which saves ground space. In addition, shopping floors are also provided for some metro hubs. The mixed development and utilization of land maximizes the economic benefits of the hub.

The billing method for the Madrid metro is similar to most European cities: subregional billing. According to the geographical characteristics of the city, the charging region is divided into six parts. At the same time, people can choose to buy tickets according to a specific time period (day ticket, monthly ticket, etc.) or the form of travel (one-way/return). Group purchase is also available. These methods create good income for Madrid metro, which is an important economic source to support the rapid development of its infrastructure.

Various intelligent elements have been added to Madrid metro operation. After the terrorist attack on the metro in 2004, Madrid metro began to adopt the advanced communication-based train control system (CBTC) to monitor and control the train operation process. The train departure interval was shortened to less than 2 minutes, which greatly increased the line capacity and ensured the safety of the trains. In addition, several lines of the Madrid metro have adopted "driverless" trains, which saves manpower and material resources by technology methods.

12.5   Suburban buses and train networks

There are more than 320 bus lines and more than 1600 vehicles operating in the suburbs, with nearly 280 million passengers transported annually [7]. The platforms of suburban buses are combined with the layout of the major urban transport transfer hub stations. The urban underground space is utilized to connect with the subway line. The transportation distance of suburban buses is two to three times that of urban buses. The distance between stations is also significantly increased, and the transportation time is also longer.

The Madrid suburban train network consists of nine lines. It is a small-scale regional railway network, independent of the national railway network, and is only operated in Madrid city. The train network is arranged in the form of a ring and radial combination, and the front platform is connected with the metro network of the interior loop public transport system. Fig. 12.9 shows the suburban train network and the subway route map of Madrid.

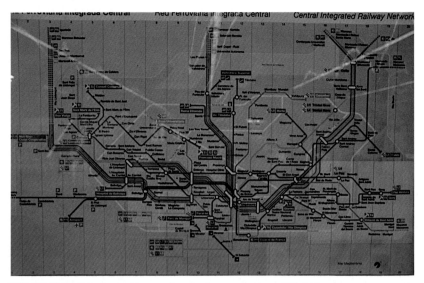

FIGURE 12.9 Suburban train network and subway route map of Madrid.

FIGURE 12.10 Long-distance run in Sun Gate Square.

12.6 Pedestrian and bicycle traffic

In downtown Madrid, people often travel on foot for short-distance trips. There are many winding alleys for people to walk through. In addition, the city government regularly organizes long-distance running to urge citizens to participate in neighborhood activities and exercise (see Fig. 12.10).

FIGURE 12.11 Cycling activities on Prado Avenue.

To promote more residents traveling by bicycle, the relevant departments of the Madrid government hold cycling activities every Sunday morning on Prado Avenue. As shown in Fig. 12.11, motor vehicles are controlled on Prado Avenue at that time. All riders wear safety helmets and children practice cycling while accompanied by adults. Cycling has thus become a popular local sport.

12.7   Madrid railway

The high-speed railway is convenient for people to travel from Madrid to other cities. Spain's high-speed railway (Renfe, shown in Fig. 12.12) is developing rapidly, constantly improving the level of railway services by high-tech means. The train runs smoothly, and there is an information service showing the temperature and speed in the carriage (Fig. 12.13). It is world class in terms of speed, ride experience, and intelligent control.

Atocha railway station in southern Madrid is an ancient building, built in 1889 with red brick exterior walls and arched roofs, as shown in Fig. 12.14. Nowadays, the scale of this railway station is far from sufficient to meet the needs of a large number of passengers, and a new comprehensive transport hub transfer center has been built close to the old railway station. The old railway station building has become a part of the comprehensive transportation hub, which has been populated with large green plants and trees. As shown in Fig. 12.15, comprehensive service facilities such as catering and shops have been set up inside the building. The station provides a warm and convenient waiting, shopping, and leisure environment for travelers.

FIGURE 12.12 Spanish high-speed train.

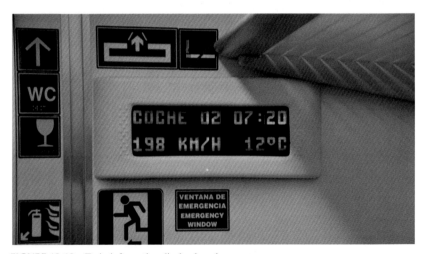

FIGURE 12.13 Train information display board.

12.8 Summary of the characteristics and traffic experience in Madrid

12.8.1 Efficient public transport system

Multiple modes of public transport in Madrid provide citizens with complete and convenient choices for public transport use. The wide coverage, good connection, regular operation, and high quality of service of the ground buses

FIGURE 12.14 Old building of Atocha railway station.

FIGURE 12.15 Warm waiting environment in Atocha railway station.

and the subway make the bus use rate in Madrid very high. In the development of its public transport, Madrid has had the following experiences:

- To guarantee bus coverage and punctuality, improve reliability;
- Intelligent monitoring and dispatching to ensure public transport safety;

- The decision-making part of the government is responsible for overall planning, and multiple operators participate in the construction management;
- The fare system has been formulated in line with the requirements of different groups of people.

12.8.2 Rapid development of railway transportation

As the capital and largest city in Spain, Madrid has become the most profitable city with the longest rail length in the process of the overall railway revival in Spain. In the process of railway construction, the Spanish government first solved people's commuting problems, and then established the national railway network based on an intelligent system. A nationwide network system with good accessibility was established by efficient analysis and expansion of the original network. This work greatly promoted the development of the railway and encouraged people to use the railway for long-distance travel.

12.8.3 Unique ancient architecture renovation

Madrid has adopted the method of "renovation" in dealing with ancient buildings. While protecting the old buildings, modern elements should be added to ensure long-term preservation and to renew their vitality in serving the people. So far, Madrid has renovated the airport, railway station, and three museums, all of which have achieved good results. The protection concept Spain has been a good reference for our country.

12.8.4 Traffic-oriented urban development

In terms of rail transit planning and design, Madrid attaches great importance to the guiding role of the rail transit system in the development of urban agglomerations and the integration of land use. All rail transit radiation lines are well integrated with the surrounding new towns, serving the need for close links between the new towns and the central cities. The design principles of transparent and seamless linkages are established in the design of transport hubs. From the different levels of the stations, the running vehicles and linkages between the upper and lower levels can be seen. A user-friendly inner environment for the metro system has been built. The government insists that the issues of seamless connection and passenger convenience must be solved, and has ensured finances and land will be available to facilitate this.

References

[1]   Wikipedia. <http://en.wikipedia.org/wiki/Madrid>.
[2]   <http://www.ctspanish.com/communities/madrid/madrid.htm>.
[3]   Wu Y. Planning and organization management of public transport system in Madrid metropolitan area. Archit Cult 2012;02:114−15.
[4]   Daniel de la HozSunchez. Mobility patterns and car use in Madrid. In: European Transport Conference; 2006.
[5]   Arnáiz M, Bueno P. Madrid: a global model for safe and efficient development of urban underground highway and railway infrastructure. Mod Tunn Technol 2009;04:1−6.
[6]   EMT de Madrid. <http://www.emtmadrid.es/Home.aspx>.
[7]   Jinjiang Z. Building urban public transport network. People's Dly 2007.

Further reading

Zheng X, Li D. Madrid metro with excellent design and easy travel. Urban Express Traffic 2012;06:112−16.
Hou J. Spanish railway policy and investment model. Railw Econ Res 2008;06 23-27 + 41.

Chapter 13

San Francisco, United States

Chapter Outline

13.1 Overview of the city

The city and county of San Francisco, also known as San Francisco, is the second largest city on the Pacific coast of the United States after Los Angeles. San Francisco is located in the northwest of California and the northern end of the San Francisco Peninsula. It faces San Francisco Bay in the east, the Pacific Ocean in the west, and has the sea on three sides. Its geographical location is shown in Fig. 13.1 [1].

San Francisco is the main city in the San Francisco Bay Area (popularly referred to as the Bay Area). It consists of four subregions: East Bay, North Bay, South Bay, and San Francisco Peninsula [2]. The topography of San Francisco and the Bay Area is shown in Fig. 13.2. It consists of nine counties (Alameda, Contra Costa, Marin, Napa, San Mateo, Santa Clara, Solano, Sonoma, and San Francisco) and 101 cities and towns. In addition to San Francisco, the main cities include Oakland in the East Bay and San Jose in the South Bay [3].

The total area of San Francisco is about 600 km², of which the land area is 121 km² and the water area is 479 km². The population is 825,000 [4], which accounts for 2.2% of California's population, the population density is about 6800 people per square kilometer [5], which is a population density in the United States only lower than New York. It is the 14th most populated city in the United States. The population, area and population density of San Francisco and Bay Area are shown in Table 13.1.

San Francisco is an important seaport city, and financial and cultural center in the western United States. It is the only administrative district in California that combines cities and counties. Surrounded by the bay waters,

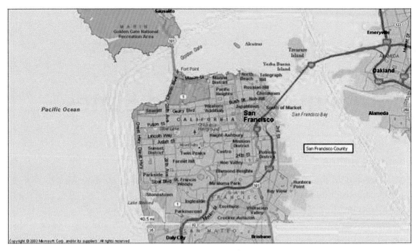

FIGURE 13.1 A sketch of the downtown area of San Francisco. *From http://relocationbreakthroughs.com/SanFranciscoMap.gif.*

FIGURE 13.2 Topographic map of San Francisco and the Bay Area.

it has beautiful scenery and a pleasant climate. The Golden Gate Bridge across the Golden Gate Strait, built in 1937, has been the largest suspension bridge in the world for 27 years. It is the landmark building of San Francisco. Therefore San Francisco also has the nicknames of the "Golden Gate City" and "Bayside City."

TABLE 13.1 Population and area statistics of San Francisco and Bay Area.

	San Francisco Bay Area	San Francisco
Land area (km^2)	10,088	121.4
Population (10,000 people)	715	82.5
Population density (people per square kilometer)	395	6800

Source: Based on Wikipedia. San Francisco, <http://en.wikipedia.org/wiki/San_Francisco>.

13.2 Urban structure and land use

Europeans arrived in the Bay Area in 1542, which in the 16th century belonged to the territory of Spain's Alta California Province and was settled by Spanish immigrants in 1776; Russia set up a post in 1806 as a material supply station for Alaska at that time. It belonged to Mexico in 1821, but was captured by the Americans in the Mexican War in 1846 and officially renamed San Francisco in 1847. At that time, it was a small town with only about 500 inhabitants. The discovery of gold deposits in the vicinity in 1848 led to a large influx of gold diggers, including the first Chinese "contract labor." When the city was established in 1850, the population had increased to 25,000, becoming a center for trade and mining services, and agriculture had developed in nearby areas. After 1869, with the passage of the transcontinental railway and the gradual improvement of port facilities, the city developed rapidly. After 1880, it began to expand to the east of the bay, forming several satellite towns. By the end of the 19th century, the population had reached 340,000. On April 18, 1906, San Francisco suffered a magnitude 8 earthquake, 80% of its buildings were destroyed, and more than 300,000 people were made homeless and forced to move to Oakland and Berkeley in the East Bay, which contributed to the development of the East Bay and its close connection with San Francisco. After the earthquake, the urban reconstruction of San Francisco formed the pattern and street view of today's modern city. In 1914, the Panama Canal was navigable, the port was booming, and trade volumes surged. In World War II, San Francisco was an important supplier of military materials. After the war, industry, commerce, finance, tourism services, and municipal construction were developed greatly. The whole Bay Area has expanded from a single central city to an urban agglomeration composed of San Francisco, the East Bay (Oakland), and San Jose [6].

San Francisco is one of the most distinctive cities in the United States. Its water area accounts for two-thirds of its total area. Its urban areas are connected by bridges with surrounding towns, and there are more than 40 hills within the city limits. The main hills are Twin Peaks, Mount Davidson, and Mount Sutro, which are above 270 m. The most famous are Nob Hill and Telegraph Hill. The central street of the city stretches from east to west and north to south in grid shape. On the basis of this grid, the road pattern fully considers the technical requirements of mountainous roads. Roads and surroundings create an abundant urban landscape and providing multi-angle viewing space for the city. Residential areas are densely built, with Market Street the busiest commercial street, extending from the city center to the northeastern corner of the city; high-rise buildings stand along Golden Gate Road; Montgomery Street and its adjacent areas are financial districts, known as "West Wall Street," where the 52-storey Bank of America Center stands. The northeastern part of the city is the main residential area. The houses there are built along hills, with the streets twisting and turning on steep slopes. The city uses a unique cable car system. Fig. 13.3 is a street view from the Bay Area to the city.

Over recent decades, the Bay Area has attached great importance to regional planning, including the establishment of public transport systems, such as Bay Area Bay Area Rapid Transit (BART) and commuter rail, the construction of Bay Bridge to connect communities across the bay, and the establishment of parks and green spaces to provide a green and healthy environment. In 2013, the Bay Area released a plan for 2040, called "Plan Bay Area." Regional planning for sustainable development was put forward. Over the past decade, the Bay Area government and local governments have worked closely to encourage employment and increase the construction of

FIGURE 13.3 Street view of San Francisco.

infrastructure, such as housing. In 2008, the Association of Bay Area Governments (ABAG) and Metropolitan Transportation Commission (MTC) proposed a plan to support local development called "Focusing Our Vision (FOCUS)," which aims to promote compact and balanced development, protect and improve the quality of life, and retain open space and agricultural resources. In recent years, this initiative has helped integrate local community development with regional land use and transport planning objectives. Local governments have identified priority development areas (PDAs) and priority conservation areas (PCAs), which form the framework of the Plan Bay Area. PDAs are usually urban central areas, urban public transport centers, etc. PCAs refer to public open areas to be protected in the long term and developed in the near future. Key PDAs and PCAs complement each other, because promoting PDAs requires closing PCAs. Fig. 13.4 shows the transportation and land uses map of the Bay Area. It can be seen from this map that land use in the Bay Area is closely related to the distribution of public transport and transportation infrastructure.

13.3   Characteristics of mobility and traffic demands

In 2010, the Bay Area had a daily average of 23.59 million trips,[1] of which 84% were by car and 5% by public transit, with 3.17 daily trips per capita. San Francisco had a daily average of 2.59 million trips, of which 53% were by car, 20% by public transit, with 3.21 daily trips per capita [8].

Fig. 13.5 reflects the proportion of residents working in other counties in each county of the Bay Area, with an average of 28% working in other counties and 72% working in the local county.[1] It can be seen that, as a whole, the Bay Area residents mainly travel within the counties, and intercounty travel traffic volume is smaller than that within the counties, which is also a manifestation of the Bay Area construction that pays great attention to job—housing balance and perfect service facilities.

Figs. 13.6—13.8 show the distribution ratios of travel distances of residents in the Bay Area under three conditions: all, noncommute, and commute. From Figs. 13.6—13.8, it can be seen that noncommute travel has a shorter distance, accounting for 42% of trips greater than 4.8 kilometers; commute travel has a longer distance, accounting for 72% of trips greater than 4.8 kilometers; all travel greater than 4.8 kilometers occupies 49%; and travel greater than 8 kilometers occupies 35%. This shows that there is a wide range of personnel exchanges among the counties in the Bay Area. Many people live in one county and work in another. The average commute distance in the Bay Area is 21 kilometers, while the average commute distance for residents in San Francisco is 11 kilometers. The commute distances of the main counties in the Bay Area (San Francisco, Alameda County,

---

1. Not including business and inter regional travel, Source: Ref 7

FIGURE 13.4 Transportation and land uses map of the San Francisco Bay Area. *From Strategy for a Sustainable Region, Draft Plan Bay Area, Association of Bay Area Governments, Metropolitan Transportation Commission, March 2013.*

Contra Costa County, and San Mateo County) are less than the average travel distances in the Bay Area, while those of remoter suburban counties are more than the average travel distances in the Bay Area.

According to 2010 statistics, 93% of households in the Bay Area have at least one car, totally more than 5 million motor vehicles and about

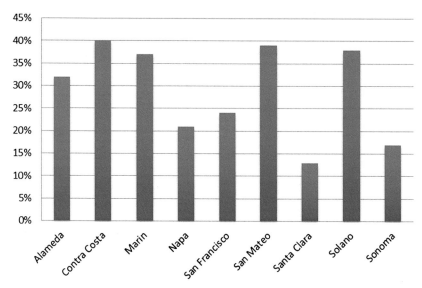

FIGURE 13.5 Proportion of residents working in other counties in the Bay Area.

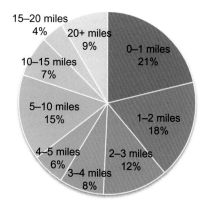

FIGURE 13.6 Distribution of all travel distance in the Bay Area.

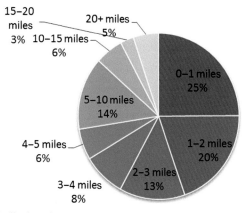

FIGURE 13.7 Distribution of noncommute travel distance in the Bay Area.

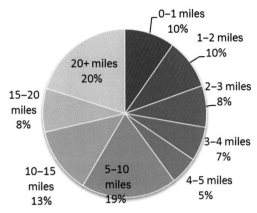

FIGURE 13.8   Distribution of commute travel distance in the Bay Area.

TABLE 13.2 Different traffic mode shares in San Francisco.

Mode	Walking	Bicycle	Public transit	Drive alone	Carpool	Total green transport
San Francisco (%)	25	2	20	53	–	47
San Francisco Bay Area (%)	10	1	5	50	34	16

Source: Based on Plan Bay Area 2040, Public Review Draft Environmental Impact Report, July 2013; San Francisco transportation update, part 2: needs assessment, <www.sfcta.org/MoveSmartSF>; Spring 2013 [9].

750 motor vehicles per 1000 people. However, these indexes in San Francisco are much smaller, where 77% of households have at least one car, with a total of 445,000 motor vehicles and about 560 motor vehicles per 1000 people. From the mode share distribution (Table 13.2), the Bay Area is still dominated by motorized travel, with cars accounting for 84% and public transit accounting for only 5%. In San Francisco, cars account for 53%, with public transit accounting for 20% and walk 25%.

13.4 Development of public transit

Public transit in the San Francisco Bay Area is quite extensive, including Bay Area Rapid Transit (BART), Caltrain, Altamont Corridor Express

FIGURE 13.9 Public transit lines in the Bay Area. *From Plan Bay Area 2040, Public Review Draft Environmental Impact Report, July 2013.*

(ACE), Amtrak, San Francisco Municipal Rail Metro (Muni Metro), Santa Clara Valley Transportation Authority Light Rail (VTA Light Rail), buses, and ferries [10]. The average daily ridership of public transit is 1.58 million. Fig. 13.9 shows the public transit lines in the Bay Area. Table 13.3 lists the main public transit services in the Bay Area.

San Francisco has a variety of buses, including cable cars, streetcars, light rail, buses, and electric buses, which are operated and managed by

TABLE 13.3 A list of public transit services in the Bay Area.

Public transit categories	Service area	Condition
Metro/heavy rail (BART)	San Francisco; Alameda County; Contra Costa County; San Mateo County; It will extend to San Jose in 2020	5 routes, 44 metro stations with a total mileage of 167 km
		Annual ridership: 123 million
		Daily ridership: 410,000
		BART is the fifth-busiest heavy rail rapid transit system in the United States[a]
		Line load intensity: 2500 person-times/km/day
		The number of passengers per hour in the late rush hour on the route to East Bay is 22,000, accounting for 89% of the carrying capacity; the number of passengers per hour in the late rush hour on the route to South Bay is 10,000, accounting for 72% of the carrying capacity
Commuter rail	Caltrain: San Francisco, Peninsula, and Silicon Valley	1 route, 29 stations with a total mileage of 46.75 miles (75 km)
		92 cars on weekdays and 32−36 cars on weekends
		Daily ridership: 40,000[b]
		Line load intensity: 500 people/km·day
		The number of passengers per hour in the late rush hour on the route to South Bay is 2400, accounting for 77% of the carrying capacity
	ACE: from San Jose to Stockton	1 route, 10 stations with a total mileage of 86 miles (138 km)
		Daily ridership: 3700[c]
		Line load intensity: 27 people/km·day

(Continued)

TABLE 13.3 (Continued)

Public transit categories	Service area	Condition
Long-distance and Intercity Rail (Amtrak)	Operation nationwide, there is only one line in the Bay Area, Capitol Corridor (from the Bay Area to Sacramento, the capital of California)	1 route, 15 stations with a total mileage of 168 miles (270 km)
		Annual ridership: 1.75 million[d]
		Line load intensity: 18 people/km · day
Light rail	Muni Metro: San Francisco	7 routes, 120 stations with a total mileage of 71.5 miles (115.1 km)
		Daily ridership: 173,000[e]
		Line load intensity: 1500 people/km · day
	VTA Light Rail: San Jose and Silicon Valley	3 routes, 62 stations with a total mileage of 42.2 miles (67.9 km) Daily ridership: 33,600[f] Line load intensity: 500 people /km · day
Bus services	There are many bus agencies in the Bay Area providing bus services, such as Muni, AC Transit, and VTA operating within the city of San Francisco, East Bay, and South Bay, respectively	
Ferries	Golden Gate Ferry: San Francisco—Larkspur and San Francisco—Sausalito, Blue and Gold Fleet and Red & White Fleet. Hornblower Cruises provides San Francisco—Alcatellas services	Golden Gate Ferry offers 3 routes and 4 stations
		Annual ridership: 2 million[g]
		The number of passengers per hour in the late rush hour on the route to East Bay is 800, accounting for 50% of the carrying capacity; the number of passengers per hour in the late rush hour on the route to North Bay is 1000, accounting for 49% of the carrying capacity

[a]Source: Ref. [11].
[b]Source: Ref. [12].
[c]Source: Ref. [13].
[d]Source: Ref. [14].
[e]Source: Ref. [15].
[f]Source: Ref. [16].
[g]Source: Ref. [17].

San Francisco Municipal Transportation Agency (SFMTA), known as San Francisco Municipal Railway (Muni Metro), which is the seventh busiest bus system in the United States. The daily ridership is 730,000 (average weekday, 2011). The average commute distance and bus travel time in San Francisco are 6.9 miles (11 km) and 44 minutes. Because of the topographic variation and the narrow streets of San Francisco, the San Francisco bus system is also the slowest mass transit system in the United States, with an average speed of only 8.1 miles/h (13 km/h). It has 1000 buses, more than 300 of which are zero-emission trolleys. There are 82 routes in the bus system, including 54 bus routes, 17 trolley routes, 7 light rail routes, 3 cable car routes, and 1 streetcar route, serving the entire city of San Francisco. The longest route is 24.1 miles (38.8 km) [18].

In order to improve the operating speed of buses, dedicated bus lanes have been set up in the main streets of San Francisco. In certain sections and periods, no vehicles other except buses and taxis, are allowed to use these lanes. By 2014, cameras have been installed in front of every bus to record vehicles entering the dedicated bus lane illegally to enable fines to be imposed. The San Francisco bus system has a wide range of services. There are bus stops in every street and intersection, with low fares, convenient transfers, and punctual buses. The bus is the most commonly used transportation method for San Francisco citizens. The travel information for the San Francisco bus system is very convenient. Passengers can check the arrival time of the next bus on by computer, mobile phone, telephone, text message, e-mail, or at the bus platform before departure. This system is called "next bus," and provides convenient and timely information for public travel.

In addition to the bus systems, it is worth mentioning the cable car system and heritage streetcars. The San Francisco cable car has a long history and was built at the beginning of the California Gold Rush. Before 1873, San Francisco relied mostly on horses for transportation, but there were more than 40 hills in San Francisco, which were undoubtedly a great test for livestock-based transportation. On August 2, 1873, the first manned cable car was opened in San Francisco, which quickly became popular with the public. During its heyday, there were about 21 routes, running 85 kilometers. After the emergence of the electric streetcar in 1892, the cable car was gradually replaced. There are still three routes in operation. Two north—south lines travel from downtown near Union Square to Fisherman's Wharf in the north, and one east—west line runs along California Street from the financial district to China Town to Nob Hill. Three lines are located following the rise and fall of the terrain. There is only one cab at the end of the cable car on the north—south line, and the cable car needs to turn around when it reaches the end point as a traditional method. As shown in Fig. 13.10, there is a turntable on the ground. The driver parks the cable car on the turntable, and then two drivers push the turntable around like a push

FIGURE 13.10   Traditional turning method for a San Francisco cable car.

mill. Nowadays, the San Francisco cable car has become one of the city's symbols. It is not only a form of public transport, but also a tourist attraction for visitors to experience the history and sightseeing of San Francisco. The cable car line follows the city's hilly terrain. It uses underground cables to pull the body forward, and the brakes and steering are operated manually. Because of its slow speed, passengers can stand on pedals of either side to see the street scenery along the way. The drivers ring a bell to remind passengers and pedestrians to pay attention before moving off, so as it travels there is a regular jingling sound, giving these cable cars the nickname "dingding cars." The crisp bell makes modern San Francisco feel somewhat old fashioned. Every July, a cable car bell-ringing contest is held by the authorities for cable car workers. With ringing bells and bright appearance, the cable cars pass through the rugged streets of modern San Francisco, creating a special scenic line. Fig. 13.11 shows a San Francisco cable car in operation.

Besides the cable car, the heritage streetcar is also a special aspect of the San Francisco bus system, as shown in Fig. 13.12. The heritage streetcar serves a line called the F Market & Wharves line, which runs 6 miles (10 km), from Castro district, the east of the Market Street, through financial districts, ferry building, to Fisherman's Wharf. This line is also the busiest one, with 720 passengers per hour in the late rush hour, which is 103% of its carrying capacity. San Francisco stipulates that the bus carrying capacity should be less than 85% of the total carrying capacity. This line obviously

FIGURE 13.11   A San Francisco cable car in operation.

FIGURE 13.12   San Francisco heritage streetcar.

exceeds this carrying capacity. The main reason for this is that the line from downtown to Fisherman's Wharf is convenient for tourists and commuters. Moreover, all the cars on this line are antique cars, and the fare is the same as that of ordinary buses, so it attracts travelers. The fact that there are many kinds of heritage streetcars used on this line also makes it stand out. Besides the streetcars that have been decommissioned in San Francisco, there are also antique vehicles collected from all over the world. These can be roughly divided into four categories [19]: President's Conference Committee (PCC) streetcars, Peter Witt streetcars, pre-PCC veteran streetcars, and a diverse collection of 10 streetcars and trams from various overseas operators. Among these, PCC streetcars from San Francisco, Philadelphia, and Newark, built between 1946 and 1948, operate on the line. Peter Witt Street Car was made in Milan, Italy, in 1928; Pre-PCC Vintage Street Car was made in 1895−1924; worldwide cars are from Britain, Germany, Japan, Australia, Russia, Portugal, and Belgium. Every car has its origin written on its body and a fixed number. They come in a variety of colors: yellow, green, orange, red, gray, and so on. They have become a typical San Francisco city scene.

13.5 Pedestrian and bicycle traffic

In addition to public transport, San Francisco has also vigorously promoted the development of nonmotorized modes of transport, such as walking and cycling. By 2008, San Francisco bicycle lanes had a distance of 335 kilometers, including 37 kilometers for Class I, 72 kilometers for Class II, 212 kilometers for Class III, and 13 kilometers of unpaved Class I. Class I, Class II, and Class III bicycle lanes are illustrated in Fig. 13.13 [20]. Class I bicycle lanes refer to independent and exclusive lanes for bicycles. Class II bicycle lanes refer to bicycle lanes marked on the same road surface as motor vehicles. Class III bicycle lanes are bicycle lanes shared with other vehicles.

In San Francisco a bicycle program was introduced in 2009, with 60 short-term (to 2014) and 24 long-term bicycle improvement plans. Fifty-two of the short-term plans involved increasing bicycle roads. A total of 55 kilometers of Class II bicycle lanes were expected to be added. Three plans were to improve intersections, two plans were to increase Class I bicycle lanes, one plan was to improve signs, and one plan was to improve Class III bicycle lanes. If the short-term plan is fully implemented within 5 years, the mileage of Class II bicycle roads would be increased by 75%. By 2011, 17.6 kilometers of Class II bicycle roads had been built [21].

In addition to increasing and improving bicycle lanes, over 1550 bicycle parking facilities have been installed by the San Francisco County Transportation Authority (SFCTA) in the past few years, with bicycle parking spaces being set up in more than 50 public parking lots, free bicycle parking spaces being set up at Caltrain stations, and passengers being allowed to carry bicycles on public transport in San Francisco—all of which contribute to improving the bicycle travel rate. According to the statistics of SFCTA in 33 locations, the number of cyclists at these locations had increased 58% from the 2006 baseline counts. About 128,000 trips are made

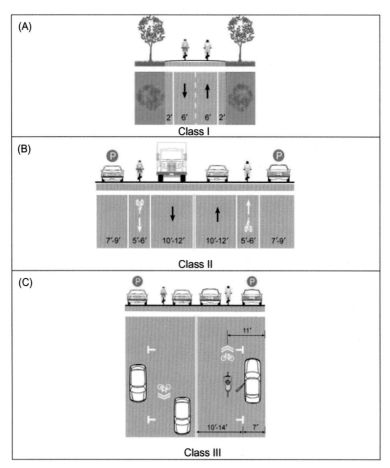

FIGURE 13.13 Schematic diagram of the classification of bicycle lanes in San Francisco. (A) Typical Class I facility—bicycle path or multi-use trail. (B) Typical Class I facility—bicycle lane. (C) Typical Class I facility—signed bicycle route. *From San Francisco Bicycle Plan, San Francisco Municipal Transportation Agency, June 2009.*

by bicycle each day in the city, or 6% of total trips [21]. The goal is to achieve 20% of total trips by 2020 [22].

San Francisco's Bike to Work Day, held in May of each year, aims to encourage commuters to try bicycling as a healthy alternative means of getting to work. On that day, many people travel to work by bicycle from different communities, and there are bicycle service stations providing water, coffee, and biscuits. On May 9, 2013, during the morning rush hour (8:30—9:30), on Market Street and Van Ness Street, 76% of total trips were made by bicycle, which set a record since 1998. In the past 5 years, Market Street surveys have shown that the proportion of bicycles has increased from 38% to 67% [23].

Since 2013, in order to reduce greenhouse gas emissions in the city, the Bay Area has launched a public bicycle-sharing system, with 70 bicycle rental stations and 700 bicycles. San Francisco has 35 bicycle rental stations and 350 bicycles. By 2014, 3000 bicycles have provided bicycle rental services for the public [22]. Bicycle stations are mostly located at the waterfront, Market Street, and railway stations to facilitate people to use bicycles for sightseeing and transfer to public transport. Long-term users have annual usage fees of $88, short-term users pay $22 every 3 days, $9 for 24 hours, $4 for 1 hour, or free for 30 minutes at a time. These fees mean that users can rent indefinitely within the service life. The urban rental stations are all networked, and bicycles can be returned to any one of the rental stations after use.

13.6   Case study: Transbay Transit Center

The original Transbay Terminal was built in the 1930s, and is a main transportation transit center in the Bay Area. Trucks and trains use the lower floor and cars use the upper floor. The designed passenger flow is 350,000 per year. Before World War II, 260,000 people took trains every year. After World War II, this number dropped to 40,000−50,000. The terminal was converted into a bus depot in 1959. In 1989, the structure was damaged in the 1989 Loma Prieta earthquake, necessitating replacement, and in 1997, the city government announced plans to rebuild it. At present, this transportation hub does meet the requirements of the current building regulations, nor can it meet the future use needs, therefore it needs comprehensive restructuring and transformation.

The new Transbay Transit Center is located in the center of San Francisco in the position of the original Transbay Terminal. It is a comprehensive transportation hub integrating rail transit (high-speed railway, railway, metro line), highway, and urban road. Transbay Transit Center is designed as an active, high-density, multipurpose area. The design reconstructs the relationship between passengers and the transit center, and integrates retail space. In sharp contrast to the original buildings, the new transit center maximizes the use of natural lighting and concentrates transit activities in a core atrium. The project was planned from 1995 to 2001, and an environmental assessment and impact analysis was conducted from 2000 to 2009, with the engineering design from 2007 to 2012, road right analysis from 2004 to 2009, construction from 2008, and opened in 2018.

The new Transbay Transit Center does not only accommodate the growing number of car journeys, but also adds train transit facilities. In the future, there will be railway linking Peninsula and East Bay and a high-speed railway linking stations farther away. The total budget of the project is about US$ 45 million, and it is planned to be completed in two phases. The first phase of construction was the construction of the aboveground bus terminal, parking garages, as well as temporary bus hubs, which was completed. The second phase of construction will include the spaced below ground and rail service, which is planned to be completed in 2025 [24]. From the completion of the transit center to 2030, it is expected that the number of passengers will reach 45,000 per year, and the number of passengers will be about 100,000 per working day; the number of commuters by railway will be 33,000 per working day; and the high-speed railway will transport 3.2 million intercity passengers and 100,000 commuter passengers per year.

The land development and utilization around the new Transbay Transit Center is very important. When the original Transbay Terminal was built in 1939, it led to the development of surrounding land which became a light industrial and commercial office-based area. After the 1989 earthquake, the connection between the Transbay Terminal and the Bay Bridge was interrupted, and the vacated land was developed into parking lots. Parking lots, offices, and bus stops remain the main land uses around the Transbay Terminal. After the completion of the new Transbay Transit Center, 2600 new residential units will be built in 40 acres (16.2 ha), 35% of which will be affordable housing (government welfare subsidized housing), 1.2 million square feet (116,000 m^2) of offices or hotels, 60,000 square feet (5580 m^2) of retail industry, and the surrounding areas will be built into a green community with parks, squares, and shaded pavements. Traveling in high-density retail business and residential areas will be dominated by public transport. The new Transbay Transit Center will increase bus convenience, reduce traffic congestion, reduce air pollution, increase the local population and employment (bringing 125,000 construction jobs and 27,000 permanent jobs), and create a prosperous business community.

The new Transbay Transit Center will form a three-in-one transportation system which integrates elevated traffic, ground traffic, and underground traffic: the two underground floors house the train hall and platform; the ground floor accommodates the bus plaza, offices, and shops; and the second floor contains the elevated bus platform, which directly connects with the Bay Bridge through a ramp. While designed to maximize the transfer of transport services, it minimizes the visual impact of the new center on the city. It is not only a transportation hub, but also a large city garden. The top floor of the center is a 5.4 acres (2.187 ha) park, spanning five blocks, including playgrounds, cafes, open-air theaters for 1000 people, and various cultural and entertainment projects, providing a leisure and entertainment venue for the public in the city center. Moreover, the 1070-foot (326-m) office, residential, and commercial high-rise building (Transbay Tower) connected to the center will be the tallest building in San Francisco [24]. Fig. 13.14A shows the elevation function distribution map of the new Transbay Transit Center and Fig. 13.14B shows the entrance map.

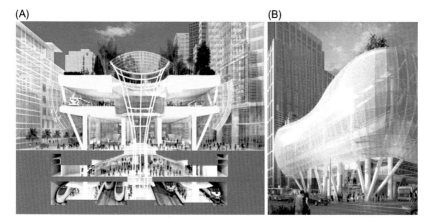

FIGURE 13.14 Distribution map (A) and entrance map (B) of the New Transbay Transit Center. *From http://transbaycenter.org/ [25].*

13.7   Summary of the traffic characteristics and experience

San Francisco is the largest financial center and an important high-technology research, design, and manufacturing base in the western United States. It is a beautiful port city and one of the 10 largest metropolitan areas in the United States. San Francisco City has a history of hundreds of years. In the process of development, great attention has been paid to the protection of the historical features of the city. In the process of new development and construction, attention has been paid to the use of advanced technologies and methods such as intelligent transportation to protect the environment and enable sustainable development, which is a good reference for the metropolises in China. The main experiences of San Francisco urban and transportation development are summarized as follows.

1. *Rational planning of land use patterns, paying attention to mixed land use and landform characteristics at different scales*

     San Francisco attaches great importance to the planning of mixed land use. The urban area is separated by Market Street. To the north are financial, commercial, and residential districts, to the south are commercial and residential districts, and to the west is a residential district. The urban area conducts compact and high-density development, plans small blocks, and considers the diversity of the community. Living services around the community are well-equipped to create a community environment suitable for green transportation. It is not only the mixed planning within the block, but also the mixed planning in the Bay Area that reduces the pressure of long-distance commuting in the city.

2. *Making use of natural landforms to create beautiful community landscape*

The hilly geography of San Francisco city means the streets must wind according to the terrain. Each street is designed according to the various wind direction, temperature, lighting, and backlight. Houses are built along the undulating terrain without destroying the landscape environment, so that each street and every building has its own style and characteristics. The famous Lombard Street is a good example of this design. It is probably the most crooked street in the world, between Hyde Street and Leavenworth Street, and about 400 m in length, and used to be a straight two-way street. Because of its steep slope, laborious and dangerous walking, in 1923, the municipal authorities transformed it into an S-shaped curved road with a single downward motorway in the middle and step-by-step sidewalks built on both sides. There are eight turns in the middle of the motorway to form an S-shaped road to graduate the slope, allowing only one-way traffic (downhill), with a speed limit of no more than 8 kilometers/h. It is paved with bricks to increase friction. In order to enhance the aesthetics of these S-shaped roads, flower beds have been built and flowers are planted in the curved blanks. Therefore this street is known as "the crookedest street in the world." Flowers and trees are planted all over the street, hydrangeas in spring, roses in summer, and chrysanthemums in autumn. Flowers and plants are cultivated in the doorways by families on both sides of the street. When the flowers blossom, from a distance, it looks like an oblique embroidery, as shown in Fig. 13.15. Lombard Street has now become a famous landscape street in San Francisco. Many tourists come here to enjoy the excitement and joy of driving on the S-slope road, while enjoying the beautiful scenery of flowers, as shown in Fig. 13.16.

3. *Giving priority to the development of public transport and maintaining different forms of public transport*

San Francisco public transport system is almost perfect, with high coverage and various forms of transport. From public transport buses, streetcars, world-famous cable cars, light rail, subway, trains, ferries, etc., it provides passengers with a variety of choices. The traveler information system is complete, providing a real-time prediction system of "next vehicle," which is convenient and attractive for travelers using public transport.

4. *Taking supporting measures (such as renting bicycles, designing pedestrian bicycle lanes, etc.) to promote the development of pedestrians and bicycles*

The San Francisco bicycle program has played an important role in promoting the development of bicycles. An improved bicycle traffic environment, as shown in Fig. 13.17, increased bicycle parking facilities and safety, and improved convenient use of bicycle rental, have encouraged

FIGURE 13.15   Looking up Lombard Street.

people to travel by bicycle every day through "San Francisco's Bike to Work Day," which has greatly improved the use of bicycles in San Francisco. In the past 5 years, Market Street surveys have shown that the proportion of bicycles has increased from 38% to 67%, with the city aiming to reach 20% by 2020. In its specific community design, San Francisco also

FIGURE 13.16 A panoramic view of Lombard Street.

FIGURE 13.17 Bicycle parking yard on Fisherman's Wharf.

pays great attention to the design of walkways, which communicate the various functions of the buildings, while extending the landscape to every corner of the community. Every year, a "Walk to Work Day" encourages people to travel on foot.

5. *Encouraging public participation in planning*

San Francisco strongly encourages public participation, such as the Association of Bay Area Governments (ABAG) and the Metropolitan Transportation Commission (MTC), which released the draft "Plan Bay Area" and arranged public hearings in nine counties of the Bay Area to collect comments on the draft. Each public hearing includes an open day scheduled for the same day, which provides Bay residents with an opportunity to watch the exhibition and ask questions about the draft plan. They encourage the public to review and comment on the draft plan online through the Plan Bay Area Town Hall, and invite residents to view the draft statutes. MTC and ABAG staff reviewed the comments and considered the adoption of the final plan for the summer of 2013 in conjunction with the reviews.

6. *Humanized urban design*

San Francisco attaches great importance to humanized design in its urban design, planning, and construction, including block design, building design, and public transport design. For example, San Francisco bus front doors are equipped with an automatic lifting platform to facilitate the disabled accessing the bus; the fronts of the buses are equipped with a bicycle frame to facilitate and encourage the transfer of bicycles with buses; street intersections are equipped with ramps to facilitate the travel of disabled people.

7. *Green infrastructure*

San Francisco is possibly the most famous green city in the world, particularly in the United States and Canada. More than 17% of the 121 km^2 of San Francisco is parks and public green spaces. San Francisco is also a leading city in green transportation. The San Francisco bus system has 1000 buses, more than 300 of which are zero-emission trolley buses. In 2011, the New York-based nonprofit Institute for Transportation & Development Policy bestowed upon San Francisco the 2011 Sustainable Transport Award for serving as a global leader in innovative polices on issues ranging from parking to bike lanes. In 2012, San Francisco won the Sustainable Transport Award from the Institute for Transportation and Development Policy. There were two major sustainable transportation initiatives. First, the development of a demand-based parking management system, with the installation of parking sensors on the roadside, the real-time provision of parking information for travelers, parking fees related to the number of parking spaces available, and real-time pricing changes to meet the demand to encourage people not to drive. Second, the city's innovative Pavement to Parks program has created new street plazas and many new parklets (sidewalk platforms that replace car parking spaces) by reclaiming street space in partnership with businesses and other community groups around the city. The parklets program has captured international attention, prompting a host of other cities to begin their own programs, from New York City to Vancouver. Fig. 13.18A shows roadside parking on Powell Street in San Francisco replaced by a sidewalk. The car in this image is the world-famous cable car. Fig. 13.18B shows Divisadero Street (see also Figs. 13.19 and 13.20).

(A)

(B)

FIGURE 13.18 San Francisco street plazas. (A) Roadside greening and famous cable car; (B) Divisadero Street greening.

FIGURE 13.19 Residential area around the Palace of Fine Arts, surrounded by green space and flowers.

FIGURE 13.20 Greening and beautification of residential surroundings.

References

[1]   <http://relocationbreakthroughs.com/SanFranciscoMap.gif>.

[2]   <http://commons.wikimedia.org/wiki/File:Bayarea_map.png>.

[3]   Wikipedia. San Francisco Bay Area, <http://en.wikipedia.org/wiki/San_Francisco_Bay_ Area>.

[4]   The California Department of Finance. E-1 population estimates for cities, counties, and the state.

[5]   Wikipedia. San Francisco, <http://en.wikipedia.org/wiki/San_Francisco>.

[6]   <http://app.travel.ifeng.com/city_intro_824>.

[7]   Strategy for a Sustainable Region, Draft Plan Bay Area, Association of Bay Area Governments, Metropolitan Transportation Commission, March 2013.

[8]   Travel Forecasts Data Summary, Transportation 2035 Plan for the San Francisco Bay Area, "Change in Motion", Metropolitan Transportation Commission; December 2008.

[9]   San Francisco transportation update, part 2: needs assessment, <www.sfcta.org/ MoveSmartSF>; 2013.

[10] Wikipedia. Transportation in the San Francisco Bay Area, <http://en.wikipedia.org/wiki/ Transportation_in_the_San_Francisco_Bay_Area>.

[11] Wikipedia. Bay Area Rapid Transit, <http://en.wikipedia.org/wiki/Bay_Area_Rapid_ Transit>.

[12] Wikipedia. Caltrain, <http://en.wikipedia.org/wiki/Caltrain>.

[13] Wikipedia. Altamont Commuter Express, <http://en.wikipedia.org/wiki/Altamont_ Commuter_Express>.

[14] Wikipedia. Amtrak, <http://en.wikipedia.org/wiki/Amtrak>.

[15] Wikipedia. Muni Metro, <http://en.wikipedia.org/wiki/Muni_Metro>.

[16] Wikipedia. Santa Clara VTA Light Rail, <http://en.wikipedia.org/wiki/Santa_Clara_ VTA_Light-rail>.

[17] Wikipedia. Golden Gate Ferry, <http://en.wikipedia.org/wiki/Golden_Gate_Ferry>.

[18] <http://www.sfmta.com/maps/categories/system-maps>.

[19] Wikipedia. San Francisco Muni, <http://en.wikipedia.org/wiki/San_Francisco_Muni>.

[20] Wikipedia. F Market&Wharves, <http://en.wikipedia.org/wiki/F_Market_%26_Wharves>.

[21] San Francisco bicycle plan, San Francisco Municipal Transportation Agency; June 2009.

[22] Wikipedia. Cycling in San Francisco, <http://en.wikipedia.org/wiki/Cycling_in_San_ Francisco>.

[23] <http://www.sfbike.org/main/today%E2%80%99s-bike-to-work-day-breaks-all-records/>.

[24] <https://www.sfmta.com/projects/salesforce-transit-center>.

[25] <http://transbaycenter.org/>.

Further reading

<http://arch.liwai.com/content-564.htm>.

Wikipedia. San Francisco Transbay development, <http://en.wikipedia.org/wiki/San_Francisco_ Transbay_development>.

Chapter 14

St. Petersburg, Russia

Chapter Outline

14.1 Overview of the city

St. Petersburg was founded in 1703 when under the leadership of Peter the Great, the Russians built a new city in a swamp. St. Petersburg was the capital for Russia for more than 200 years. In 1914, it was renamed "Petrograd," and after Lenin's death in 1924, it was renamed "Leningrad." After the collapse of the Soviet Union in 1991, the original name "St. Petersburg" was restored. Today, St. Petersburg is the second largest

Eco-Cities and Green Transport. DOI: https://doi.org/10.1016/B978-0-12-821516-6.00014-X

291

city in Russia and its most important port city. St. Petersburg, is also a famous historical and cultural city in Russia, with a magnificent palace, the Summer Palace, which is a resplendent and magnificent solemn church, known as the "Russian Paltiel agriculture temple." The navy museum was once Russia's tallest building, Paul fortress, a noble and elegant small monastery, etc. With many fine buildings, the whole city is a akin to a large building museum. Unlike Moscow, St. Petersburg is a purely European-style city, and a "window to Europe" for Russia.

St. Petersburg is situated on the eastern bank of Finland Bay in the Baltic Sea, where the Greater Neva River and the Lesser Neva River converge. It is a "water city." The city consists of more than 40 islands, with more than 70 natural rivers and canals circling between them. The water system in the city is very developed (Fig. 14.1), with the rivers running vertically and horizontally, with sparkling blue water and elegant architecture. There are more than 300 ancient and modern bridges. These are truly bridge museums and have given St. Petersburg the reputation of "Venice in the North."

St. Petersburg consists of 20 districts, 16 prefectural municipalities, 10 district municipalities, 41 towns, and 204 administrative farms [1]. The city covers an area of 1439 km^2, with a population of 5.03 million and an average population density of 3440 people per square kilometer [2]. Fig. 14.2 shows a bird's eye view of the Cape of Vasily Island, where the Neva river is divided into the Greater Neva river and the Lesser Neva river, and continues to flow toward the Gulf of Finland.

FIGURE 14.1 Map of St. Petersburg.

FIGURE 14.2   An aerial view of the St. Petersburg Vasily Island headland. *From http://zh.wiki-pedia.org/wiki/File:St_petersburg_metro_map_sb_zh-hans.svg.*

14.2   Urban structure and land use

St. Petersburg covers an area of 1439 km², and is 44 km from north to south and 25 km from east to west. It has an irregular radial pattern centered on the winter palace. The city is divided into four parts by the Greater Neva river and Lesser Neva river into the Moscow, Vasily Island, Petrograd, and Viborg districts. The left bank of the Neva river (south bank) is the Moscow district, Vasily island district is located between the big and the small Neva river; the east of the big Neva river is the Vyborgshy district, which is connected by more than 300 bridges. St. Petersburg city forms a huge horseshoe-shaped town group at the eastern end of Finland Bay (including Catlin Island): the north bank is a residential area, health resort, and tourist area, while the south bank is a cultural and commercial center; the Neva river stretches as far as Ivanovskoye in the east [3].

As a famous historical and cultural city, St. Petersburg's urban planning has its own characteristics. St. Petersburg is a city built entirely according to plan since its founding in 1703. Bloom put forward the famous "Ideal Plan" in 1717, which is the strategic development plan of the capital located on the Neva river. This unified the island into the oval city surrounded by a canal for the first time forming its own unique cultural tone [4].

After World War II, the city was mainly developed to the southwest and along the coast, and new streets like Engels, Moscow, and Shomilov, as well as seaside parks and coastal streets, were built. In the overall urban planning scheme of 2005, although the city has been expanded a great deal, it maintains the old city's road network form, original shape, and spatial scale of streets [4].

FIGURE 14.3 Winter Palace Square in St. Petersburg.

The scholars and well-known experts involved in the urban planning of St. Petersburg have also absorbed the opinions of its citizens. The main goal is to create a comfortable urban environment and improve the quality of life of the citizens.

Today, streets in St. Petersburg are centered on the Winter Palace Square (Fig. 14.3), with the main trunk of Nevsky Avenue, showing an irregular radial shape. However, the roads parallel to or perpendicular to the main street are arranged in a checkerboard shape, which is a European-style urban road design. Generally, there are four lanes in both directions. Parking is allowed on the road according to the traffic flow, and parking alternately is allowed at a 45-degree angle in sections with wide roads and low traffic volume [5].

St. Petersburg has many unique cultural buildings. It is the largest historical center in Europe, covering an area of 3400 ha. The historic center is a national treasure, and an "open-air museum," which is included in the UNESCO World Heritage List. In the process of determining the pattern and scale of St. Petersburg's urban construction, great importance has been attached to the protection of these historical relics. To this end, the government has promulgated a series of laws to protect the historical and cultural heritage, which impose strict restrictions on tall and new buildings. All construction projects in the old urban areas have to be tested using 3D models. Through the special planning for the protection and development of the historic central area, the municipal government ensures the protection of the cultural heritage, the renewal of engineering and road infrastructure, the restoration of buildings, and the improvement of parks, while improving the quality of buildings around the central area. To date, there are no skyscrapers in the old city of St. Petersburg. The streets and buildings which are over 300 years old remain unchanged. Even new buildings inherit the original style, form, color, height, and other aspects of the older buildings.

14.3 Development of public transit

St. Petersburg has developed convenient transportation. Subways, buses, taxis, urban railways, and water transportation constitute the urban public transport network of St. Petersburg.

14.3.1 Subway

St. Petersburg has five subway lines, 110.2 km long, totaling 58 stations, including 6 transfer stations. Sixty percent of St. Petersburg travelers use the subway system, which has a daily passenger traffic of 3 million people. It is the main means of transportation for the citizens of St. Petersburg [6].

Signs in the subway system are clear; there are relief arrows on the walls of each station to indicate the direction of the subway direction on the platform. In the subway transfer station, the hall is divided into two or three floors, mostly using "T" or "parallel" transfer mode, and the transfer distance is not more than 100 m, also, passengers transfer through overpasses or underground passages, which is very convenient.

Most of the stations are classic in design, elegant in materials, and rich in artistic atmosphere. Each subway station is like an underground palace, with luxurious buildings and different styles. All kinds of reliefs and frescoes are inlaid with colorful marble, granite, ceramics, and stained glass, supplemented by fantastic lighting fixtures. From ceiling to floor, from painting to sculpture, the works are novel in conception, delicate in production, and varied in genres. No design pattern in a station is repeated. During this visits, people may feel as if they are in an art palace [7].

The metro in St. Petersburg is punctual, fast, and cheap. The departure interval of subway trains is 2−3 minutes during the day, and 5 minutes at night [8]. Subway stations are also equipped with electronic clocks that show the time between the arrival and departure of two trains, allowing passengers to monitor whether the trains are on time.

14.3.2 Buses

St. Petersburg has a developed public transport system, including buses, trolleys, and trams. There are 113 bus lines, 55 tram lines and 46 trolleybus lines, with a total of more than 4300 transport vehicles [8]. Vehicles on the urban central line are relatively new, fares are unified at 2 roubles, and the departure interval is short, which basically makes up for all areas that are not be covered by the metro.

St. Petersburg has the world's second largest trolley-bus system after Moscow. The symbol of trams usually start with M, and the symbol of most buses start with A.

Bus route instructions are clear, with all vehicles on the same road having their stop signs combined. If you are going to turn to another road ahead, the stop signs are placed separately. Therefore as long as you look at a stop sign, you can be assured that this line of vehicles will reach the next station.

14.3.3 Water transport

St. Petersburg waterway passenger transport is very seasonal, its main source of custom is tourists, who use it for sightseeing along the river. The most famous routes are from the Winter Palace to the Summer Palace by hydrofoil yacht and the city canal sightseeing route [9]. Fig. 14.4 shows a passenger ship on the Neva river.

St. Petersburg is Russia's most important land and water transport hub, is Russia's gateway to the rest of the world, and is its largest seaport (Figs. 14.5 and 14.6). There are 12 railways radiating to major cities such as Moscow and that are also connected with railways in Poland and Finland. The main port area is located in the southwest of the city. It has more than 50 berths and can berth seagoing vessels with draft water of about 11 m. Its annual cargo throughput is 10 million tons. It is connected with regular flights to Nordic and western Europe. The harbor freezes in winter, requiring

FIGURE 14.4 Passenger ships on the Neva river.

FIGURE 14.5   "Venice in the North"—St. Petersburg: overlooking Vasily Island and fortress.

FIGURE 14.6   The Nikolai bridge opens at 2 p.m. and 11 p.m. daily.

icebreakers. St. Petersburg is also one of the largest river ports in Russia. It is the starting point of the Baltic−White Sea Canal and the Baltic−Volga Canal. It is also the estuary of the Neva river, and has well-developed river transportation. Ships can reach the White Sea, the Volga river,

FIGURE 14.7   Hiring a bicycle in St. Petersburg. *From http://www.flickr.com/photos/bluehat-fyx/8169338041/sizes/o/in/photostream/.*

the Caspian Sea, the Black Sea, and the Azov Sea from here. St. Petersburg also has an important airport, with more than 60 routes leading to cities at home and abroad. The highway traffic is also well-developed, with 11 highways converging here [10].

14.4   Bicycle traffic

St. Petersburg is flat and suitable for cycling (Fig. 14.7). In 2008, the Mayor of St. Petersburg decided to enthusiastically build bicycle lanes and bicycle ring routes. He formulated a 3-year bicycle development plan, which stipulated that bicycle lanes should be built in urban trunk lines, gardens, and parks, and at the same time opened up special training routes for children. Some bike lanes were planned, such as line one, line three and line seven for cyclists in Neva district. The municipal government not only calls for the establishment of bicycle lanes in the districts, but also plans to require municipal officials to ride bicycles to and from work after the completion of the bicycle lanes [11]. Today, St. Petersburg has implemented the Freebike program enabling bicycles to be used free of charge.

14.5   Summary of the traffic characteristics and experience

After more than 300 years, St. Petersburg still retains its charm from the beginning of its construction and follows its original planning (Figs. 14.8−14.26). In the process of traffic development, St. Petersburg

FIGURE 14.8   Panoramic view of St. Petersburg Summer Palace.

FIGURE 14.9   Summer Palace. Looking along the axis of the Summer Palace garden is the Bay of Finland. The largest characteristic of this design is the integration of artificial landscape and natural landscape. A close view of the artificial canal and the magnificent view of the Gulf of Finland.

FIGURE 14.10 The sculpture in the trapezoidal waterfall of the Summer Palace.

FIGURE 14.11 Mushroom fountain in the Summer Palace: one of the features of the Summer Palace is a variety of forms and colorful fountains. These fountains are characterized by individual features.

FIGURE 14.12 Bouquet fountain in the Summer Palace.

FIGURE 14.13 Dragon waterfall in the Summer Palace.

advocates a green travel mode without destroying the traditional style, historic buildings, and urban landscape of the city. It integrates all kinds of traffic modes well and is convenient and harmonious.

FIGURE 14.14 The upper garden of the Summer Palace emphasizes solitude and elegance, providing a quiet space for reflection and relaxation.

FIGURE 14.15 Sidewalks, landscape, and seats for visitors to rest in St. Petersburg's Neva Street.

FIGURE 14.16 The arcade building on Neva Street can protect visitors from rain in summer and cold in winter, providing leisure and shopping convenience for visitors.

FIGURE 14.17 Sloped road underpass, avoiding the need for stairs.

FIGURE 14.18 The pedestrian street on both sides of Neva Street: Pleasant architectural scale, appropriate street facility and store settings to meet people's daily needs and providing convenience and comfort for travelers, tourists, and people at leisure.

FIGURE 14.19 Department of history, University of Petersburg: arcade design with a safe and comfortable walking space.

FIGURE 14.20   Coffee and tea bar on Neva Street.

FIGURE 14.21   Bleeding Cathedral in St. Petersburg.

In conclusion, St. Petersburg has the following characteristics in terms of transportation and urban landscape:

1. Development under the guidance of planning. Planning ideas and urban development concepts at the beginning of the city construction have been inherited and developed through more than 300 years.

FIGURE 14.22 The church of Isaac and the statue of Nicholas I.

FIGURE 14.23 The sphinx of the Neva river as an architectural sculpture adorns the landscape along the river. The landscape zone along the river provides a platform for people to stop, take photos, and enjoy the water.

FIGURE 14.24 Cameron cloister at Yekaterina Palace.

FIGURE 14.25 Exchange building on Vasily Island: this building is called the Parthenon on Vasily Island. The creative metal model of the front square is both a sign and an orientation guide.

FIGURE 14.26 Art plaza: the appropriate space layout and the perfect rest facilities for visitors reflect subtle humanistic care.

2. Only by building top-quality infrastructure can we create a top-quality city. St. Petersburg attaches great importance to the architectural monolithic and street landscape, the magnificent Summer Palace, the solemn and elegant Winter Palace, the broad and comfortable Neva Street, and the rich and distinctive traditional streets, all of which constitute the unique temperament and character of St. Petersburg. St. Petersburg has a strong artistic atmosphere, good livable environment, and distinct historical and cultural characteristics, and green transportation is dominant. The whole city is in harmony with nature.

3. Road design fully guarantees pedestrian space, and also pays attention to the setting of pedestrian rest facilities and landscape viewing needs, with roadside coffee shops etc., which add to the city's interest and facilitate the needs of travelers, making the city very welcoming and people-friendly.

4. Giving priority to the development of public transport. Although St. Petersburg's population density is not high and its economy is well developed, the public transport system remains the main mode of transportation for citizens. In addition, the city government encourages bicycle travel and is moving toward a new level of green transportation. At a time of rapid urban development in China, St. Petersburg's urban experience has provided a very valuable lesson for similar ancient capitals and historical and cultural cities. How to enable the city to develop a new vitality while maintaining its traditions is a subject we should think deeply about.

References

[1]   <http://www.baike.com/wiki/%E5%9C%A3%E5%BD%BC%E5%BE%97%E5%A0%A1>.

[2]   <http://zh.wikipedia.org/wiki/%E5%9C%A3%E5%BD%BC%E5%BE%97%E5%A0%A1>.

[3]   <http://city.cri.cn/29344/2011/03/01/4207s2560299.htm>.

[4]   Yan W, Ma J. Study on the characteristics and protection experience of the historic and cultural city St. Petersburg. Res Urban Dev 2012;7.

[5]   Yu T. A trip to St. Petersburg. Traffic transp 2007;.

[6]   <http://zh.wikipedia.org/wiki/%E5%9C%A3%E5%BD%BC%E5%BE%97%E5%A0%A1%E5%9C%B0%E9%93%81>.

[7]   Wan X. Talk about the experience of visiting the Russian subway. Urban Transp 2003;.

[8]   <http://shengbidebao.h.baike.com/article-191089.html>.

[9]   <http://www.mofcom.gov.cn/article/zt_shanglvfw/lanmufour/201310/20131000361522.shtml>.

[10]   <http://euroasia.cass.cn/news/105460.htm>.

[11]   <http://news.sdinfo.net/gjxw/429952.shtml>.

Chapter 15

Rio de Janeiro, Brazil

Chapter Outline

15.1 Introduction

Rio de Janeiro is Brazil's second largest city and its largest seaport. It has a well-developed business and financial industry. Because of the wonderful scenery there, it is also a famous tourist city. Rio de Janeiro is located in the Gulf of Guanabara in southeastern Brazil, with the Atlantic Ocean on its south side. The city of Rio de Janeiro covers an area of 1256 km^2 and has a population of 6.32 million. In January 1502, the Portuguese explorer Pedro Alvarez Cabral arrived, mistaking the Guanabara Bay as an estuary of a large river, he named it Rio de Janeiro (Portuguese for "January River"). In 1555, the French Navy established the first European colony here, but in 1565, the Portuguese invaded Rio de Janeiro, expelling the French and occupying the city. To defend against the French and pirates, the Portuguese began building castles, starting with the original "Rio de Janeiro." In 1808, Rio de Janeiro served as the capital of the Portuguese empire and housed its nobles, and in 1822, Pedro I declared independence for Brazil and Rio de Janeiro was established as the capital. On April 21, 1960, the Brazilian capital was moved to Brasilia. Although it is no longer the capital of Brazil, Rio de Janeiro is still the most distinctive and beautiful city in Brazil and possibly even in South America.

Corcovado is 710 m above sea level and is located in the heart of the city of Rio de Janeiro. The 38-m-high statue of Jesus, which was completed in 1931 on the top of the mountain, is the landmark of Rio de Janeiro and was selected as one of the new Seven Wonders of the World in 2007. Unlike the traditional image of Jesus, he does not have a pained expression in this statue. For more than 80 years, Jesus, with a flat face and kind face, opened his wide arms and looked down and embraced this beautiful city.

Eco-Cities and Green Transport. DOI: https://doi.org/10.1016/B978-0-12-821516-6.00015-1

FIGURE 15.1 Overlooking the city and bay from the top of Kokowado mountain.

When visitors climb to the top of Corcovado and stand by the image of Jesus, they can see the beautiful scenery: the high and low urban buildings and the reefs in Guanabara Bay form a unique landscape. The sun is scattered on the sparkling sea, and the city glitters like a fairyland in the glow of water. The 400-m-high hemispherical rock close to sea level between the Atlantic Ocean and the Guanabara Bay is nicknamed Sugarloaf Mountain (Fig. 15.1) and is another landmark in Rio de Janeiro.

Taking the cable car to the top of the sugar loaf, facing the Atlantic Ocean, on the right is a curved new moon-shaped piece of land, 4 km long Kopakapana (Fig. 15.2), and on the left is the semicircular Botafogo beach (Fig. 15.3), with a wonderful night view under the lights.

15.2 Promenade

Rio de Janeiro has a coastline of more than 600 km and is a natural place to enjoy the sea shore, with its natural white sand beach. When Rio de Janeiro first built the city road, it was organically integrated into the natural landscape. Between the famous Copacabana beach and the Atlantic Avenue (Av. Atlantica), and between Ibanema beach and Av. Francisco, there is only a low curbstone. As shown in Fig. 15.4, from left to right are motorways, bicycle lanes, sidewalks, and beaches. People can enjoy the Atlantic waves and white beaches as they travel.

Along the scenic coastal boulevard, a continuous, flat two-way bicycle path and a corrugated pattern of sidewalks are designed to create ample space between the road and the beach, thereby providing people with a good

FIGURE 15.2   Kopakapana beach.

FIGURE 15.3   Nightscape of Botafogo beach.

environment for recreation. On both sides of the road there are urban cultural landscapes with high-rise buildings, and on the other side, blue sea and white sand beaches. People can enjoy the gentle sea breeze and warm sunshine while enjoying the beautiful view of the modern city and nature.

FIGURE 15.4 Atlantic Avenue along Kopakapana beach.

15.3 Public transit

Rio de Janeiro has a well-developed public transport system with more than 440 bus lines and carries about 4 million passengers a day. On the streets of the city center, buses are everywhere, and there are very few private cars (Fig. 15.5). There are special bus lanes on the roads in the city which are clearly signed (Fig. 15.6). There are several large bus hubs in the city (Fig. 15.7), from which buses from multiple lines are issued.

15.4 Bicycle rental system

Rio de Janeiro has a 74-km bicycle-only road, and there are a bicycle rental sites at the central isolation zone of the motor vehicle roads (Fig. 15.8). Citizens can rent bicycles using their IC card. Many citizens choose to ride a bicycle with the family on holidays to relax or exercise (Fig. 15.9).

15.5 Solving traffic problems for the "slums"

From the 1950s to the 1990s, Brazil's urbanization rate was too fast, causing a series of urban problems such as poor housing, unemployment, public insecurity, and inadequate sewage and waste disposal. The most prominent problem was that a large number of poor people built on the hillsides within the urban area. Simple homes, dense, with a lack of infrastructure, form poor settlements, known as "slums." Rio de Janeiro and São Paulo are the two most prominent cities in Brazil. There are currently more than 700 slums of

FIGURE 15.5   Buses running on the streets of downtown Rio de Janeiro.

FIGURE 15.6   Bus lanes in Rio de Janeiro.

different sizes in Rio de Janeiro. According to the 2010 census, about 22% of the poor live in these slums. This is a painful lesson in the process of urbanization for Brazil that took place during the second half of the 20th century. It will take a long time and a lot of effort to solve this issue.

(A)

(B)

FIGURE 15.7 (A) Entrance to a bus hub station in Rio de Janeiro. (B) The neatly arranged departure platform in a bus hub station hub in Rio de Janeiro.

The dilapidated houses in the slum areas are dense and the roads are narrow. Most of the houses are built on steep hillsides, and access is extremely inconvenient. This also brings a lot of security and environmental problems

FIGURE 15.8   Bicycle rental point, Rio de Janeiro.

FIGURE 15.9   Cyclists in Rio de Janeiro.

(Figs. 15.10 and 15.11). In order to solve the residents' travel and environmental problems, the government started by solving the problem of basic transportation facilities and began building a cable car system to reach the poor inhabitants on the mountain top (Fig. 15.12).

FIGURE 15.10 Slums in Rio de Janeiro on the mid-mountain.

FIGURE 15.11 Rundown houses, and narrow and steep roads in slums.

15.6 People-friendly walkways

The design of the walkways in Rio de Janeiro is people-friendly. Most of the pavement is paved with black and white stones in the traditional rich

FIGURE 15.12   Construction of a cable car to a hilltop slum.

FIGURE 15.13   A Portuguese-style grid pedestrian road.

Portuguese style. In the center of the city, many sidewalks are laid out in the form of a grid as shown in Fig. 15.13, while the sidewalks on the beaches are laid in waves, appearing to be an extension of the waves to the land (Fig. 15.14).

FIGURE 15.14 Paved wavy pavement along the coastline.

15.7 Romantic stairs

In the LAPA area of Rio de Janeiro, there is a small, romantic, and environmentally friendly trail—the Escadaria Selaron. It was originally a normal stepped road with a width of about 5 m and about 250 steps. The buildings and walls on both sides are very worn, as shown in Fig. 15.15. In 1990, the Chilean civilian artist Selaron decorated the front of the stairs and the walls on both sides with tile fragments recovered from a waste dump. He collected landscape pictures of famous cities and Brazilian map patterns made of broken tiles (Fig. 15.16). It has now become an important part of the landscape in Rio de Janeiro (Fig. 15.17), with many tourists visiting daily. They are able to enjoy the author's careful design, the pursuit of environmental protection, and the Rio de Janeiro people's tolerance for multiculturalism and respect for unofficial art.

15.8 Summary

The greatest feature of Rio de Janeiro is the harmonious integration of the city and the environment. In this city, people tend to be friendly by nature. In urban construction, designers have paid attention to the integration of artificial buildings and natural landscapes. In the planning and construction of

FIGURE 15.15   Escadaria Selaron staircase in Rio de Janeiro (top-down).

FIGURE 15.16   Ceramic tile veneers with various patterns of the Escadaria Selaron staircase.

FIGURE 15.17 Escadaria Selaron stepped overall landscape (from bottom to top).

transportation facilities, planners make good use of superior natural conditions and combine artificial facilities with natural beauty. At the same time, public transportation and green transportation systems, such as bicycles and walking, have become the mainstream of urban travel.

Chapter 16

Carmel, United States

Chapter Outline

Carmel is a quiet, romantic, and artistic town located on the west coast of the United States, at the northern end of the Monterey Peninsula. The town was founded in 1903, and has a sandy beach and vast nearby forest. The natural environment is superior, attracting many poets, writers, artists, actors, and other celebrities from the United States to live there. The town currently has more than 4000 inhabitants.

16.1 Harmony with nature

For more than 100 years, although the United States has been world leader in modern methods, the residents of Carmel have always adhered to the principles of nature and original ecology. There are no high-rise buildings, no karaoke dance halls, no flashing neon lights, and there are no traffic lights on the roads. A small number of cars are quietly driving on clean streets, with pedestrians taking precedence. The town's central area is only about 1000 m long and is located in a stone arch on the central square of the town, giving a natural and quaint impression (Fig. 16.1). One of the main streets, Ocean Ave., runs through the central square and heads west to Carmel beach. The pine trees on the shoreline sway in the direction of the town with a long-term sea breeze (Fig. 16.2).

The residential buildings lined up along the coastline are nestled among the old towering trees, which are beautifully shaped and varied in style. In Fig. 16.3 is a stone building, with the facade of a natural stone house, like a European castle; in Fig. 16.4 is a house with a very simple design of wooden materials, without any modification; and in Fig. 16.5 is a modern house with

Eco-Cities and Green Transport. DOI: https://doi.org/10.1016/B978-0-12-821516-6.00016-3
323

FIGURE 16.1 Central plaza and stone arch in Carmel.

FIGURE 16.2 Trees slanting toward town under the long-term sea breeze.

light weight, and elegant color and style. These buildings are integrated with the blue sky and white sand and green trees facing each other, fully demonstrating the harmony and beauty of the human landscape and the natural scenery.

FIGURE 16.3   Stone buildings on the shore of Carmel beach.

FIGURE 16.4   A pristine wooden house on the shore of Carmel beach.

16.2   Landscaping and building beautification

The residents of Carmel pay great attention to the greening and beautification of their blocks and buildings. The buildings here are not luxurious, most are simple and unpretentious, but they use natural trees and flowers to

FIGURE 16.5 A lightweight and stylish house on the shore of Carmel beach.

FIGURE 16.6 Flowers blooming around buildings.

decorate the environment and create a step-by-step view of the town. As shown in Fig. 16.6, the simple wooden structure of a residential building is surrounded by flowers, like walking into the Garden of Eden. Fig. 16.7

FIGURE 16.7   Greening and beautification of a building wall façade.

shows the three-dimensional greening and beautification of the façade of a building. Walking on the sidewalk next to such a building is like being completely in the natural environment.

The buildings in Fig. 16.8 are common in color and shape, but plant a green plant in a narrow passageway of the building and trim it into a special shape with lavender herbs and flowers on the ground, and it is transformed. Combined with the white buildings, it is no longer a bland home, but constitutes a wonderful picture. All of these show the exquisite intentions of the residents of the town and the relentless pursuit of beauty.

Fig. 16.9 is a landscaping design for the corner of a street. From the apex of the intersection of the two roads, it is stepped upwards, planting flowers step by step, and the top is a tall green tree, which ensures that the intersection has a good view and expands. The landscaping area forms an unfolded three-dimensional beautification effect.

The roads in the town are relatively narrow, with trees on both sides. The residents here are also quite unique in the pruning of their trees. As shown in Fig. 16.10, the trees are carefully trimmed and shaped to create a deep and fresh artistic effect on the roads in the neighborhood.

In addition to the use of flowers and plants for landscaping, the town also uses examples of exquisite sculptures to enhance the art of the town. Fig. 16.11 is a bronze statue standing at the entrance to a shopping street. The lively figure and the smooth lines mean that pedestrians look at the

FIGURE 16.8 The perfect combination of architecture and trees.

FIGURE 16.9 Street corner expansion stereo beautification design.

statue as if they can hear the beautiful music she is playing. A bronze sculpture adds vitality to the place. On the way to the town, what you can see and hear can be said to be works of art, and Carmen is well-deservedly called an "art town."

FIGURE 16.10   The artistic effect of carefully pruned trees to create streets.

FIGURE 16.11   Statue at the entrance to a shopping street.

16.3   Simple facilities and signs

Carmel is not only an art town, but also a town that is people-oriented and thoughtfully serving its residents and tourists. On the streets of the town,

FIGURE 16.12   Public mailbox set on the street.

you can see some facilities and signs that are convenient for residents and tourists. In the public letter box shown in Fig. 16.12, residents and visitors are free to access information. The public restroom shown in Fig. 16.13 is indicated by the signpost for the convenience of travelers.

In front of some shops, on both sides of the road, and in public places such as shopping streets, there are seats and umbrellas for visitors to rest at (Figs. 16.14 and 16.15). The bus stop is also very simple, but practical (Fig. 16.16).

16.4 Summary

Carmel makes full use of the advantages of its coastal natural scenery to create a perfect landscape town. The town's overall elegance and taste are refreshing and memorable.

1. Each building unit is very sophisticated, with distinctive personal features and creative architectural design; the single-family home gardens next to the coast have their own unique features, and constitute a unique landscape of the whole street.
2. The star-studded family hotels and street-facing shops are small in scale, but pay special attention to beautification and have a warm homely feeling, so that visitors can relax and enjoy them fully.

FIGURE 16.13   Signs for public toilets.

FIGURE 16.14   Tables and chairs set in front of a shop for visitors to rest.

3. The whole town is full of people-friendly design. There are seats for people to rest along the street, in front of the stores, and around the family
   hotels. The signs for the public toilets are clear, and the bus stops are
   equipped with chairs for passengers to rest while waiting.

FIGURE 16.15 Chairs and umbrellas in a shopping street.

FIGURE 16.16 A simple bus stop.

4. The landscape development around the town is dominated by the natural state, with appropriate human intervention and guidance, fully demonstrating and practicing the ecological concept and the better realm of harmony between humans and nature.

5. There is no traffic light signaling throughout the town, which keeps the street quiet.

Carmel does not pursue luxury. It respects nature, integrates nature, pursues harmony with nature, and uses elegant art and unpretentious techniques to make Carmel a hugely charming and perfect ecological town. The experience of Carmel is worth learning from when building a boutique town.

Chapter 17

Letchworth, United Kingdom

Chapter Outline

17.1 Howard's garden city theory

17.1.1 Introduction to Howard and the background to his theory

Ebenezer Howard (1855−1928) was born in a commoner family in London, England [1]. During his youth, he began to learn about the works and ideas of Walt Whitman (1819−92) [2], Ralph Waldo Emerson (1803−82) [3], and others. He was most influenced by *The Age of Reason* [4] written by Thomas Paine (1737−1809), a famous statesman in the United States [5]. During his time in America, Ebenezer Howard also read some books about urban planning, for example, *Hygeia* written by Benjamin Ward Richardson (1828−96) [6]. After returning to England, he worked for a long time as a stenographer in parliament. Through his work, he was in contact with a large number of political, economic, and social officials and scholars, who gave him have a deeper understanding of the problems existing in the community organization, urban and rural structure, as well as population distribution existing in the social system of Britain at that time.

In the 1760s, with the start of the Industrial Revolution in Britain, mankind moved from an agricultural civilization to an industrial civilization. The Industrial Revolution brought fundamental changes to the modes of production and social form of people. Machine factories replaced manual workshops, and production efficiency was greatly improved. A large number of rural people flocked to cities, promoting the urbanization process. With the process of industrialization, cities expanded rapidly and the population also

Eco-Cities and Green Transport. DOI: https://doi.org/10.1016/B978-0-12-821516-6.00017-5

increased at a tremendous rate. As a result, there was a housing shortage, environmental degradation, and other urban problems were exacerbated. Problems such as excessive population agglomeration, environmental pollution, and the serious gap between rich and poor grew in London, Manchester, New York, Chicago, and other metropolises. Many urban problems aroused the attention and debate of politicians and sociologists, and the future of urban development was a conundrum to most at that time. Under this background, the London government carried out a great deal of research to find suitable solutions.

Ebenezer Howard was authorized with the investigation and analysis of urban problems and their solutions. Influenced by the thoughts of the British social reform movement, he carried out a thorough investigation and thinking aiming at a variety of social problems, such as land ownership, tax revenue, urban poverty, urban expansion, and deterioration of the living environment, and gradually developed a comprehensive understanding of the kinds of problems facing the development of an integrated urban and rural framework for society [7]. In October 1898, he published *Tomorrow: A Peaceful Path to Real Reform* [8], in which he proposed schemes for building novel cities. In 1902, the second edition of the book was published, and the title was changed to *Garden Cities of To-morrow*. Up to now, six editions have been published and issued in Britain.

Ebenezer Howard's garden city plan is not simply a study from the perspective of urban structure and layout of the city, but rather through comprehensive planning of urban space, and establishing independent urban management institutions and investment−return mechanisms to seek the integration and coordinated development of urban and rural areas. The essence of the book is a declaration of social reform under the social background at that time. Howard's garden city theory is an important theoretical base for the British garden city movement. In 1902, the planning, designate, and establishment process of the world's first garden city, Letchworth, was started. This was also the successful practice case of Howard's garden city. In addition to the two garden cities, Letchworth and Welwyn, built in Britain, they also appeared in Austria, Australia, Belgium, France, Germany, the Netherlands, Poland, Russia, Spain, and the United States [9].

17.1.2 Concept and design key points of garden city theory

In a broad sense, Howard's garden city theory is not just a kind of form or city planning on drawing, in essence it is a kind of social reform, that is to make a city scape and the overall planning of urban organisms, which constituted the society and played a key role in the development of society. It not only includes the overall layout of the city and its relationship with the surrounding cities, but also a large part of the planning for the organisms that affect the development of the city, so as to solve the problems caused by the continuous abnormal development of the metropolis and prevent the uncontrolled development of the metropolis. This theory has great qualities, and is described as the beginning of modern urban planning.

The purpose of the garden city theory is to build a city that is beneficial to the life and healthy living of its citizens, with reasonable layout of suitable residential, commercial, and industrial buildings. The scale is not too large, but it should be able to meet the needs of all kinds of social life, with a population size of approximately 32,000, and covering an area of about 9000 acres. The city is surrounded by farmland, which produces agricultural products that can sold to those living in the city. All the land is owned by the public or entrusted to the community. The center of the city is a mixed zone of commercial areas, housing areas, and industrial areas, creating a city with economic independence.

The core concept and design goal of the garden city is to enable people to live in the city with both good social and economic conditions and a beautiful and natural ecological environment. The most obvious feature is to emphasize the harmony and unity of urban and rural areas, and the city should reflect the characteristics of beauty of nature, social justice, and urban−rural integration.

17.2   Overview of the garden city of Letchworth

Letchworth was the world's first garden city, and was created by Howard himself. It is located 38 miles north of London. Two monuments stand in the city streets, where they are planted with green grass and blooming flowers, as shown in Figs. 17.1 and 17.2. These are "Welcome To The World's First Garden City" and "Ebenezer Howard founded the town in 1903," telling

FIGURE 17.1   The first garden city monument.

FIGURE 17.2 Letchworth garden city monument created by Howard.

visitors the story of the founder of the garden city more than 100 years ago in silence.

17.2.1 Historical evolution

The history of the Letchworth garden city is as follows [7].

In 1899, Howard organized and established the Garden City Association to publicize his theories and ideas of garden cities.

In 1902, the Garden City Association decided to put Howard's theory into practice and build the first garden city, as a social experiment to see if Howard's idea worked.

In 1903, they raised £300,000 as their first start-up capital. In Hertfordshire, 38 miles north of London, 3818 acres of Letchworth land was purchased and the first "Garden City Co., Ltd." was created to be responsible for the planning, design, construction, and other specific work of this garden city. Architect Barry Peck and planner Raymond Unwin were commissioned to create a master plan for the idyllic city. This was the official start of this great social experiment.

In October 1903, at the opening ceremony of the project, after digging the first sod, Earl Grey said: "I think we should congratulate Mr. Ebenezer Howard for his vision for the future will be realized after five years. Lucky community residents will be delighted to hear the news:

The natural appreciation of the rents of thirty thousand people would not go to the pockets of some landlord, but will increase investment to improve the material and spiritual lives of all residents."

In 1904, the master plan for Letchworth was drawn up, based on Howard's garden city theory and design principles, with the general layout of the city in a grid pattern. To divide the city functions, the industry zones are concentrated in one area, the residence zones are arranged in the periphery of the industrial zones, and the sheltered forests and boulevards are planted between the industrial zones and the residence zones. The business district is in the center of the city, with the train station located in the center of the city and the industrial zone. The brownstone districts are located on the edge of the city, and configurations of mixed communities are also taken into account.

Most of the houses in the garden city have large gardens, surrounded by green space, the elaborate building layout making full use of sunlight to create a natural and warm living environment. The roads, trees, hedgerows, and large areas of green land are under elaborate landscape planning and convey the rural life into the city. In 1905, the first phase of the planning and construction was completed. The early residences were relatively uniform. After decades of continuous construction until the 1970s, the garden city was completed. The development of Letchworth soon demonstrated the many advantages of the garden city concept and became a model for the construction of garden cities in Europe and around the world. The early residences of Barry and Unwin became the standard for many similar constructions.

17.2.2   Present situation of the city

At present, the total area of Letchworth garden city has reached 22.26 km^2, with a population of 33,600. It is administration area at the town level in Hertfordshire, England. Under the background of rapid improvement of modernization and urbanization, 9 km^2 of farmland still remains. There are 780 enterprises providing 15,100 employment positions, accounting for nearly half the population. The construction area of the employment area is 0.39 km^2, social housing accounts for 33%, and the maximum residential density is 3000 households/km^2. There are 22 km of garden green roads within the scope of the city.

Since the beginning of the 21st century, the appearance of Letchworth garden city has been changing quietly. In industrial areas, new types of industry are gradually increasing. The pace of innovation in residential and commercial areas is accelerating, new supermarkets replace country groceries, and new plans are being made for business centers. However, residents still try to maintain the characteristics of a garden city, continue to uphold the original idea, keeping the esthetics principle in the buildings and urban surroundings, creating a comfortable garden city, and providing a good living and working environment for the residents. The museum records the history of this garden city, and the heritage foundation working as an organization gives a detailed introduction to Howard's ideas and the practical experience of the city for each visitor, adhering to the pursuit of the ideal livable environment of the garden city.

17.2.3 Architectural and landscape design characteristics

17.2.3.1 Rustic cottage architecture

Letchworth garden city is built in the traditional village architecture style. In its early days, ordinary residents adopted the design of high vault, pediment, and row style (as shown in Fig. 17.3). Roof windows were set on sloping roofs, and the simple small garden in front of each house added a pastoral effect to the house.

17.2.3.2 Houses standing in the middle of a garden

As shown in Fig. 17.4, wildflowers bloom on the wide grass in front of a building, fully showing the beautiful natural scenery. In Figs. 17.5 and 17.6, gardens and lawns are surrounded by flowers and trees. Fig. 17.7 shows residential buildings which have been constructed recently. They have adopt the enclosed design, with many families sharing the green space.

17.2.3.3 Pursuit of the quiet rural life

Letchworth pursued a simple, natural rural life, abandoning a bustling and noisy life. Fig. 17.8 shows the first coffee shop in Letchworth. To create the quiet atmosphere of a garden city, early Letchworth decided not to build public entertainment facilities. Along with the increase in the number of residents, in order to create a place for gatherings and necessary communication by the people, by a vote of all citizens, a simple cafe shop was built. It is in the simple architectural style and has an elegant and quiet environment.

17.2.3.4 Pedestrian and leisure spaces set in the central block

Fig. 17.9 shows the central block of Letchworth garden city. There is a spacious leisure space between the buildings and the road, with fountains and

FIGURE 17.3 Early townhouse.

FIGURE 17.4 Wide grassy area in front of a residential building.

FIGURE 17.5 Residential buildings shaded by flowers and trees.

sculptures to increase the quality of life of citizens. Fig. 17.10 shows the downtown shopping street in the center of Letchworth, which is equipped with a welcoming walking space and leisure facilities.

FIGURE 17.6 Manicured lawns and hedges.

FIGURE 17.7 Enclosed residential buildings, with a number of families sharing the green space.

17.2.3.5 Construction of green tree-lined avenues

From the beginning of its construction, Letchworth garden city has paid special attention to the construction of green tree-lined avenues. After more than 100 years of construction, the city is now covered with trees and grass,

FIGURE 17.8   The first coffee shop in Letchworth.

FIGURE 17.9   Leisure spaces set in the central block.

with a total length of 22 km of tree-lined avenues (Figs. 17.11−17.13), providing comfortable and fresh walking spaces for citizens.

17.2.3.6   Public building landscape

The building shown in Fig. 17.14 was originally a textile factory workshop. After the factory was abandoned, it was transformed into an office building, with a water feature in front of the building, with trees and flowers planted around it, to green and beautify the office environment and public space.

FIGURE 17.10 Downtown shopping street in Letchworth garden city.

FIGURE 17.11 Current situation of tree-lined avenues in Letchworth garden city.

17.3 Inspiration from Letchworth

More than 100 years ago, the wave of urbanization and rapid urban expansion brought about by the Industrial Revolution caused environmental pollution, the deterioration of public security, and other urban problems. Howard

FIGURE 17.12   Current situation of tree-lined avenues in Letchworth garden city.

FIGURE 17.13   Current situation of tree-lined avenues in Letchworth garden city.

FIGURE 17.14   A public building transformed from an abandoned factory workshop.

put forward the theory of a garden city, trying to build a livable city with a modern industrial civilization and beautiful rural scenery. The creation of Letchworth garden city is a successful outcome of Howard's theory.

Today, the scientific and technological level of mankind has reached a new level, with the global population now about four times that of a century ago, significantly larger than at the time Howard was alive. However, the coordinated growth between nature and humanity, the environmental protection, the creation of a warm and livable urban environment, as well as the idea of building a fair and harmonious society which were put forward by the garden city theory is still a goal we pursue with unremitting efforts.

From a technical point of view, the garden city proposed by Howard can be seen as a basic unit constituting the modern magacity.

References

[1] Wikipedia. Ebenezer Howard, <http://zh.wikipedia.org/wiki/Ebenezer_Howard>.
[2] Wikipedia. Walt Whitman, <http://zh.wikipedia.org/wiki/Walt_Whitman>.
[3] Wikipedia. Ralph Waldo Emerson, <http://zh.wikipedia.org/wiki/Ralph_Waldo_Emerson>.
[4] Wikipedia. Thomas Paine, <http://zh.wikipedia.org/wiki/Thomas_Paine>.
[5] Paine T. In: Tian F, Xu W, editors. The age of reason. Beijing: China Legal Publishing House; 2011.
[6] Wikipedia. Benjamin Ward Richardson, <http://en.wikipedia.org/wiki/Benjamin_Richardson>.
[7] Barton Willmore. New Garden City. London: Barton Willmore; 2012.
[8] Howard E. Garden Cities of tomorrow. Trans. Jin jing yuan, Beijing: Commercial Press; 2010.
[9] Baidu Baike. Garden cities of tomorrow, <http://baike.baidu.com>.

Further reading

Howard E. Tomorrow: a peaceful path to real reform. London: S. Sonnenschein & Co., Ltd.; 1902.

Chapter 18

Malmo, Sweden

Chapter Outline

18.1 Introduction

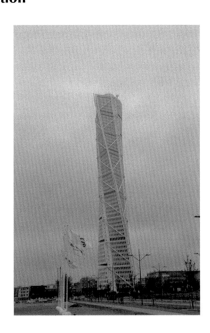

Malmo is the third largest city in Sweden, located in the southern Swedish province of Skona, on the east bank of the Erle Strait of the Baltic Sea, across the sea from Copenhagen, the capital of Denmark. The 16-km long Olson bridge built in 2000 connects the two cities across the sea, and is a link between northern Europe and the European continent. The total population of Malmo is about 292,000, the urban built-up area is 156 km^2, and the population density is about 1872 person/km^2.

Malmo is built along the sea shore, and the entire city presents a semicircle shape, surrounded by two ring roads. Originally, Malmo was a port and industrial city. However, with the decline of the shipping industry, Malmo successfully transformed into a comfortable, livable, low-carbon, and energy-saving ecological city through a series of high-quality planning and construction projects. At the same time, Malmo has also made remarkable achievements in the construction of a green transportation system. The number of motor vehicles in Malmo is about 108,000, but the share of walking and bicycle travel has reached 43%.

18.2 Case study of the Western Harbor

The Western Harbor is located in the northwest of Malmo, with an area of 175 ha. The Western Harbor was previously a shipbuilding industrial base, with tower cranes and water edge completely occupied by industrial facilities such as docks and material warehouses. Since 1998, the industrial facilities have been dismantled and rebuilt. Now, a green and beautiful community has been formed, which is a low-carbon mixed type, environmental and ecological community. The Western Harbor is a successful model of transformation from a pure industrial land to a high-quality residential, convenient life and travel, ecological livable community with comprehensive functions. This success comes from an advanced idea, with mixed use and development of land, perfect planning and low-carbon energy-saving and environmental protection. It has an important influence on future land reconstruction and new city construction.

18.2.1 The goal of the Western Harbor plan

The development and construction of the Western Harbor began in 1998. Before construction, the planning department made a comprehensive and detailed plan for the area, population, number of houses, transportation mode, roads, energy, commercial facilities, waste and sewage treatment system, and park green spaces, and determined that the Western Harbor should be jointly developed by multiple developers (Table 18.1).

Compared with Malmo, the planned population density of the Western Harbor is higher, and the land use is diversified according to the mixed development mode, including residential buildings, office buildings, commercial facilities, schools, green parks, etc. The biggest development

TABLE 18.1 Planning contents and objectives of Western Harbor.

Western Harbor	
Population and land	
Population	4326
Households	2558
Planning population	20,000
Planning jobs	17,000
Planning area	175 ha
Planned population density	11,428 persons/km^2
Transportation	
Car ownership per 1000 people	440
Mode split	Walking: 31%
	Bicycle: 29%
	Public transportation: 17%
	Private car: 23% (among them Multiple Occupancy Vehicle 3%)
Project situation	
Development and construction started in 1998 and is expected to be completed in 2021. It is jointly developed by several developers.	
Four schools and 15 kindergartens are planned.	

highlight is to pay attention to low-carbon environmental protection, energy conservation and emission reduction, and ecological livability (Fig. 18.1). This is mainly reflected in the following aspects:

1. The electricity mainly comes from wind energy generated by an offshore wind power plant;
2. Solar panels and heat pump systems are mainly used for heating;
3. Recycling of water resources. There is a channel along the wall of each house to guide the rainwater and gather it to the center of the community to form a natural small pond;
4. Waste recycling. In 2008, 95.8% of the waste was classified for recycling or converted into biogas, with only 4.2% of waste going to landfill;
5. Almost all building roofs are afforested, which is beautiful and can absorb rainwater to prevent flooding and overheat. In summer, air conditioning is generally not needed.

FIGURE 18.1 Community panorama of Western Harbor.

18.2.2 The traffic system

The traffic in the West Port is very convenient, and green traffic is the main mode. At present, the proportion of walking and cycling has reached 60%. This is mainly due to the attention paid to the balance of jobs and residences in the Western Harbor, and the relatively self-sufficient facilities such as residence, work, school, and business. There are kindergartens and schools in the community. One-third of the residents can go shopping within 500 m, and 37% of the residents' commuting distance is less than 5 km [1]. The effective implementation of these plans greatly reduces the long-distance journeys of the residents in the community.

The traffic system design of the Western Harbor is considered from the perspective of environmental protection, with promotion of green traffic modes such as bicycle, walking, and public transportation. The motor vehicle driving route should be set at the periphery of the community as far as possible, while the bicycle and pedestrian roads should be built inside the community. In the community there are 10 bus stops.

A good bicycle traffic system has been designed for the Western Harbor (Fig. 18.2). There are clear bicycle roads in the community, and a bicycle signal phase is provided. A large bicycle parking lot (Fig. 18.3) is set up near the bus station and the railway station to facilitate the parking and transfer of travelers and to promote the traffic mode of bike + train or bus. A bicycle rental system has been set up to facilitate its use and operates at a

FIGURE 18.2   Bicycle lanes in the Western Harbor.

FIGURE 18.3   Bicycle parking near the bus stop.

low price. These measures encourage people to use bicycles in the Western Harbor. Almost every household owns bicycles, and the share of bicycle travel has reached 29%.

18.2.3 The blocks of the Western Harbor

The Western Harbor is divided into 10 blocks. At present, seven blocks have been developed and constructed, and the remaining three blocks, Mashusen, Dockan, and Harmporten, will be gradually developed in the next few years.

18.2.3.1 Block Bo 01

Block Bo 01 is located in the northwest of the Western Harbor, covering an area of 22 ha, and was the first block to be developed. At present, there are 1394 households, with about 2293 residents, and also commercial and office buildings and a parking building. As the first part of a new community, block Bo 01 has realized the utilization of 100% renewable energy and maintains diversity in building forms.

Of course, as the first exploratory new community, there have also been some problems in the development of Bo 01 block. The most prominent problem has been that the standard orientation of the community building is relatively high, leading to the gradual evolution of a community in which the rich people live together. Therefore, the number of motor vehicles in this community is relatively high, with an average of more than one vehicle per household, which was not the original goal of the Western Harbor. Although a very convenient pedestrian and bicycle road has been designed in the community, it is rarely used. Parking is a big problem, and an additional parking building has been built.

18.2.3.2 Flagghusen block

Flagghusen block is located in the north of the Western Harbor, covering an area of 4 ha, with 16 buildings and about 600 houses, two-thirds of which are public rental houses. In the process of construction, great attention was paid to the diversity of housing, including providing affordable housing for low- and middle-income citizens. At the same time, attention was paid to the scale of the buildings in line with people's views, and so the height of all buildings is under 25 m. In terms of energy conservation, the residential design stipulates that the energy consumption per square meter shall not exceed 120 kW.

18.2.3.3 Fullriggaren

Fullriggaren is located in the north of the Western Harbor, and currently has 634 apartments, 75% of which are public rental. Similarly, considering the mixed use and development of the land, office buildings, a kindergarten and a park were built in the block. The biggest highlight of the development of Fullriggaren block is passive energy-saving buildings and environmental protection facilities. In the area, waste is collected through independent pipelines to produce biogas, realizing the resource-based utilization of waste.

FIGURE 18.4   Central green park in the Western Harbor.

Solar panels and wind power generation provide part of the renewable energy used in the area. The wall greening and roof greening of buildings is a major feature of the regional architecture.

18.2.3.4   Motesplatser

The Motesplatser block, also known as Varvsparken, is located in the center of the Western Harbor, where the largest central park of the Western Harbor (Figs. 18.4 and 18.5) has been built, including a theme park, a place for children to play, entertainment facilities for residents' leisure and fitness, and a separate development park for pet dogs (Fig. 18.6).

The park has been put into construction since the beginning of the development of the Western Harbor, to ensure that the green landscape area of this plot is not occupied by subsequent projects, and also greatly improves the landscape effect and living quality of surrounding buildings, so that residents have a perfect green park and leisure and entertainment place from the beginning of occupancy. On the west side of the central park is a school (Figs. 18.7 and 18.8), which provides convenience for children in the community to go to school nearby.

18.2.3.5   Kappseglaren

Kappseglaren block is located in the middle of the Western Harbor and the east side of the central park. There are 320 houses planned here, of which about 220 are public rental houses with a residential area of about

FIGURE 18.5   Entertainment facilities in central park in the Western Harbor.

FIGURE 18.6   Pet dog park in the Western Harbor.

30,000 m². In addition, there are 14,000 m² of office buildings, and a small community park (koggparken) has been built in the northwestern part of the block. The block is a high-density mixed development plot, which fully shows the diversity of building types and land use. In addition, passive

FIGURE 18.7   School on the west side of the central park.

FIGURE 18.8   Talking with children in a school in the Western Harbor.

building energy-saving and low-carbon technologies are used to minimize energy consumption and carbon emissions through the whole life cycle of building construction and use.

FIGURE 18.9   Historical buildings near the water in the Western Harbor.

18.2.3.6   Varvsstaden

Varvsstaden area is located in the south of the central part of the Western Harbor, with 1500 households planned and about 5000 jobs provided. It used to be an industrial park, but has been transformed into a high-tech industrial park. Dozens of enterprises have settled here, providing a large number of jobs for the Western Harbor. In this area, there is the largest inner lake in the Western Harbor, with many historical buildings near the water (Fig. 18.9). At the same time, riverside space is used to build houses (Fig. 18.10) to create an ecological living environment and to beautify the landscape for community residents. At present, the plot is still under development.

18.2.3.7   Universitetsholmen

Universitetsholmen block is located in the south of the Western Harbor, and is the block closest to Malmo city center. The environment here is elegant, with a beautiful waterfront landscape and it is close to the downtown area, therefore many universities are gathered here. With the rise of the university town, some large-scale conference centers, hotels, concert halls, and other cultural and entertainment facilities are also settled here. At the same time, it also attracted some high-tech industries and office buildings, including Mercedes Benz and other famous companies, forming a high-tech industrial park (Fig. 18.11). In order to reduce the energy consumption of these large-scale public buildings, efforts have been made in building materials, lighting design, and other aspects by using building energy-saving technology. For

FIGURE 18.10 Riverside space in the Western Harbor.

FIGURE 18.11 High-tech industrial park in the Western Harbor.

example, extensive use of glass materials (Fig. 18.12) and an increase in natural lighting areas can reduce the energy consumption of artificial lighting. The energy consumption of buildings in this area is 25% less than that specified in the national standard of Sweden.

FIGURE 18.12 Glass walls to increase the natural lighting areas of buildings.

After more than 10 years of development and construction, the Western Harbor today has become a new type of community successfully transformed after deindustrialization. The building type and land use are mixed, which is conducive to the balance of work and residence. In terms of transportation system design, green transportation modes such as bicycles, walking, and public transportation are promoted. In terms of energy conservation and environmental protection, building energy conservation, low-carbon emission technology, and the design concept of "passive building," using renewable energies such as wind energy, solar energy, and trash recycling have been adopted to create an energy-saving and environment-friendly ecological community. In terms of living environment and landscape design, green parks and recreational spaces are set up to create a welcoming ecological living environment with varied building shapes, which is a successful case of ecological community development (Figs. 18.13−18.17).

18.3 Successful experience

As a medium-sized city, Malmo is a very important inspiration and reference for a large number of small and medium-sized cities and towns to build livable, environment-friendly and energy-saving eco-cities. Based on the prospective concept and planning of ecological livability, environmental protection, and energy conservation, the development mode of mixed land

FIGURE 18.13   Residential area and street park in the Western Harbor.

FIGURE 18.14   Living environment near water in the Western Harbor.

use and practical and effective construction and implementation, the Western Harbor has become a model of ecological city construction. The main experiences are as follows.

FIGURE 18.15 Landscaping around residential buildings in the Western Harbor.

FIGURE 18.16 Diversified architectural forms in the Western Harbor.

18.3.1 Good urban planning

In the 20th century, with the unprecedented achievements of scientific and technological progress, we have entered a new modern society. While

FIGURE 18.17   Natural ecological landscape in the Western Harbor.

enjoying the material civilization brought by high technology, people have found that the human consumption rate of resources has reached epidemic proportions, and the damage to the natural environment has also reached a most serious level. Therefore, saving energy, reducing pollution, protecting the environment, and returning to nature have become a higher goal of the 21st century. The planning and construction of the Western Harbor is based on this concept. Moreover, the planning and design of the Western Harbor is carried out with the participation of the public, and each building is determined through competition. Every building has a detailed planning control to ensure coordination with the whole community.

18.3.2   Mixed land use and green transport

Residential buildings, shops, office buildings, schools, and green parks are mixed through each block of the Western Harbor. Residents can easily work, shop, go to school, and enjoy leisure activities nearby, which greatly reduces the demand for long-distance travel and improves the proportion of walking and cycling. At present, the share of bicycle travel in the Western Harbor is 29%.

18.3.3   Mixed living of different types of people

At the beginning of the planning for the area, the Western Harbor was considered as having the characteristics of mixed living of different types of people. According to different income groups, it was planned for houses of different grades, different types, and different natures, including expensive individual houses and a large number of relatively cheap public rental houses, to protect the Western Harbor from becoming a community inhabited only by the rich. A mixed living community is conducive to the exchange and understanding of different income groups, increasing social integration and vitality, and also, to a certain extent, reducing the community's car ownership rate.

18.3.4 Design of buildings and streets

The Western Harbor prefers new shapes and uniqueness in the design of buildings. Each building selects the best design scheme through a competition. Only the winner of the competition has the opportunity to obtain the land development right. After winning the bid, construction must be carried out in full accordance with the design scheme, so that the buildings in the community have their own characteristics and the overall style is coordinated and unified. The unique twisted shape of a turning torso in block Bo 01 makes it a landmark building of the Western Harbor and even Malmo.

18.3.5 Ecological community

The planning goal of the Western Harbor is to make full use of renewable energy, minimize waste discharge, recycle natural rainwater resources, and build an ecological community. After more than 10 years of development and construction, the Western Harbor has become an ecological city with green transportation. It can be a global example in the creation of an idyllic environment, combined with the characteristics of natural geography to build a high-grade ecological community, which has achieved the planning goal and become a model ecological community. many measures have been implemented such as wind power generation and solar energy, waste sorting and recycling, and biogas production using waste. A large number of buildings have become "passive buildings" through natural lighting, ventilation design, and the use of thermal insulation materials. Therefore, the Western Harbor realizes the ecological community vision of energy saving and low-carbon emissions.

Reference

[1] Fletta N, Henderson J. Low Car(bon) Communities: Inspiring car-free and car-lite urban futures. 2016.

Further reading

Wikipedia. Malmo, <http://zh.wikipedia.org/zh-cn/>.
Current Urban Development in Western Harbor; 2012.
Xi-li H, Sjostrom P (Sweden). Landscape architecture in the sustainable urban development: Bo 01 eco-community in Western Harbour in Malmo, Sweden. Landsc Archit, 2011(4). [in Chinese].

Chapter 19

San Carlos, Brazil

Chapter Outline

19.1 Introduction

San Carlos is a small inland city, about 200 km northwest of São Paulo in Brazil. It is rarely known to outsiders and has few visitors. The built-up area is 49.385 km^2. In 2013, the population was 236,457 and the population density was 4788 per km^2. In June 2013, the total number of motor vehicles was 151,742, including 97,802 cars, 26,390 motorcycles, and 3644 small-displacement motorcycles (scooters). The shares of different modes of transportation are shown in Table 19.1. This is a small city with fresh air and clean streets. The north to south Av. San Carlos is the main street in this city. One-way roads have been constructed that are the width of two motor vehicles and they have parking spaces on the side. There are neatly arranged public buildings such as shops, coffee shops, churches, kindergartens, and schools (Fig. 19.1).

Fig. 19.2 is a cathedral located in the center of the city. The white façade is covered with green trees, and the pale-yellow domes create a solemn and sacred atmosphere.

The grounds to the church square contain a city pattern of San Carlos in a ratio of 1:200 paved with black and white stones (Fig. 19.3). Although it has been worn out over time and it has now become unclear, it still illustrates the meticulous attention of its urban designers.

19.2 Color and green design of block buildings

Brazil is described as a developing countries, and San Carlos is a small city with an underdeveloped economy. There are no high-end buildings and buildings are not highly decorated, but residents here pay great attention to the use of color changes and greening and beautification design to make their homes more lively (Fig. 19.4). The designers flexibly use various colors on

Eco-Cities and Green Transport. DOI: https://doi.org/10.1016/B978-0-12-821516-6.00019-9
363

TABLE 19.1 Transportation share of San Carlos.

Transportation	Walking	Cycling	Public bus	Private bus	Others
Share (%)	29	3	19	37	12
Green transportation mode share (%)	51				

Note: Data are provided by Professor Nielsen, University of Sao Paulo, Carlos.

FIGURE 19.1 San Carlos Avenue.

the façades of ordinary buildings to make the street landscape more varied and interesting. As shown in Figs. 19.5 and 19.6, the green plants and artificial buildings are organically combined, and the walls are vertically greened to fill the street landscape. This vitality reflects the harmonious beauty of architecture and the natural environment. The clean and tidy streets are unpretentious and give a peaceful impression.

19.3 Building an ecological city

Constrained by its level of economic development, the urban infrastructure of San Carlos is extremely basic, some buildings are in need of repair, and many wires and cables are exposed, which adversely affects the

FIGURE 19.2 Cathedral building.

FIGURE 19.3 Map of San Carlos on the ground of St. Carlos Cathedral Square.

environmental landscape, however people here are still able to pay attention
to the natural ecology. The preservation of space creates a welcoming,
practical, and peaceful living environment under the simple conditions.
Along San Carlos Avenue, there are parks of about 1 ha every 200–300 m.

(A)

(B)

FIGURE 19.4 (A—D) Changes in the colors of the façades of ordinary buildings.

These parks are completely open and there are no fences. The parks are covered with trees and flowers, and there are no deliberate artificial modifications (Figs. 19.7 and 19.8). The parks provide quiet walking spaces for residents and are also good places for citizens to enjoy their leisure time. Some sculptures, water fountains, and other landscape pieces (Fig. 19.9) have been placed on both sides of the street to make it more interesting. Pedestrians can also take a short break and relax here. Fig. 19.10 shows a

(C)

(D)

FIGURE 19.4   (Continued).

pedestrianized shopping street with some simple green plants, creating a welcoming and friendly environment for casual shopping. Fig. 19.11 shows a bus stop, which is simple but neat, giving a plain impression.

19.4   Summary

Compared with Carmel in the United States, San Carlos has an underdeveloped economy. Streets and buildings are still slowly being modernized, with

FIGURE 19.5   Beautifying effect of the organic combination of green plants and buildings.

FIGURE 19.6   Vertical greening of walls fills buildings with vitality.

exposed aerial wires, and the urban construction is far from perfect. However, this does not affect the pursuit of a better environment for people in developing countries. People here use simple methods to beautify their homes. In the process of modernization, the precious ecological environment has been protected. Decent and unique single buildings, comfortable and

FIGURE 19.7   Street park with natural ecology.

FIGURE 19.8   Street park with natural ecology.

elegant shopping districts, flowery church squares, and streetscapes that blend in with the natural environment give an impression of comfort and pleasure. This case is useful and instructive for the construction of small cities in China's economically underdeveloped regions.

FIGURE 19.9 There is leisure space on both sides of the road.

FIGURE 19.10 Green space on a pedestrianized shopping street.

FIGURE 19.11   A simple and practical bus station.

Chapter 20

Los Angeles, United States

Chapter Outline

20.1 Overview of the city

Los Angeles is located on the west coast of the United States, in Southern California. Los Angeles city has a population of 3.858 million (2012), an urban area of 1214 km^2, and a population density of 3178 people/km^2. Los Angeles city, together with its surrounding long beach and other cities, constitutes the metropolitan area of Los Angeles, with a total area of about 12,519.6 km^2, a population of about 12.828 million, and a population density of 1024.6 people/km^2 (Table 20.1).

Eco-Cities and Green Transport. DOI: https://doi.org/10.1016/B978-0-12-821516-6.00020-5
373

TABLE 20.1 Profile of population and area of Los Angeles.

	Los Angeles metropolitan area	Los Angeles city
Area (km^2)	12,519.6	1214
Population (millions)	12.828	3.858
Density (person/km^2)	1024.6	3178

20.2 Urban structure and land use

Los Angeles has always been regarded as a typical case of suburbanization and urban sprawl, and is the most representative city with disorderly urban sprawl and car-dominated transportation in the United States. With the development of its population and social economy, the urban area of Los Angeles has greatly expanded. Before 1900, the urban area of Los Angeles changed very slowly, but between 1900 and 1940, the urban area increased more than 10-fold. In 1915, San Fernando Valley was incorporated into Los Angeles, which increased the urban area by 440 km^2.

The geographical span of Los Angeles Metropolitan Area is more than 100 km in east–west and north–south directions. Less than 6% of the city's population is employed in downtown Los Angeles. As a result, the importance of downtown Los Angeles is relatively low compared with other cities of the same size in the United States. In Los Angeles, there are about 10 subcenters in addition to the city center. Outside Los Angeles, there are 16 cities with populations of more than 100,000, and 29 cities with populations between 50,000 and 100,000. Therefore, the multicenter, low-density, and horizontal urban spatial structure of Los Angeles is very obvious, and is in sharp contrast with the single-center, high-density, and vertical urban spatial structure of New York.

Its urban sprawl makes Los Angeles continue to expand to its surrounding suburbs, and urban residents gradually move from urban residential areas (mainly apartments) to single villas in the suburbs, far away from the city center. The density of land use is declining, as residents move far away from their work place and have to use private cars to travel to work. As a result, urban traffic is highly dependent on cars, causing air pollution (including particulate pollution and greenhouse gas emissions), traffic jams, and high energy consumption. Los Angeles's current traffic structure and problems are closely related to its unlimited urban sprawl. Fig. 20.1 shows a panoramic view of Los Angeles.

20.3 Motorization and traffic demands

Los Angeles is famous for its "car culture," with only 16.53% of households not owning cars (54.5% in New York). The number of motor vehicles in

FIGURE 20.1   A panoramic view of Los Angeles. *From Wikipedia: Los Angeles. http://zh. wikipedia.org/.*

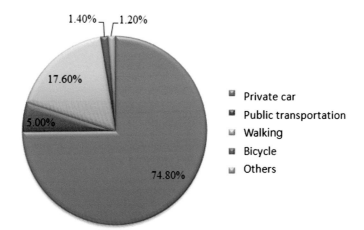

Private car
Public transportation
Walking
Bicycle
Others

FIGURE 20.2   Travel mode split of Los Angeles. *From Travel in LA County by Nancy McGuckin, 2009.*

Los Angeles Metropolitan Area is about 6.43 million, and the number of motor vehicles per 1000 people is 540. Fig. 20.2 shows the spatial distribution of the per capita number of motor vehicles in Los Angeles. It can be seen from this figure that, except for the central area, there are about 600 vehicles for 1000 people, and more than 800 vehicles for 1000 people in some areas (source: Density, Car Ownership, and What It Means for the Future of Los Angeles).

TABLE 20.2 Travel mode split of Los Angeles.

Mode	Public transportation (%)	Walking (%)	Bicycles (%)	Private cars (%)	Others (%)	Green transport in total (%)	Year
Los Angeles city	5.0	17.6	1.4	74.8	1.2	24.0	2009
Los Angeles metropolitan area	2.5	10.2		87.3		12.7	2012

FIGURE 20.3 Elevated light rail in downtown Los Angeles.

Los Angeles residents travel 3.52 times a day, including 2.63 times a day by car. The average travel distance is 10.3 km, with the average travel distance by car is 13 km. In terms of the choice of transportation mode, private cars also play a leading role. Three-fourths of trips are completed by cars, while public transportation only accounts for 5% (see Table 20.2 and Fig. 20.2).

20.4 Public transportation

Fig. 20.3 shows the elevated light rail in Los Angeles, and Fig. 20.4 shows the distinctive Hollywood subway station. In Los Angeles, the decoration

FIGURE 20.4   Distinctive Hollywood subway station platform.

style of many subway stations reflects the industrial or commercial characteristics of the site, which are very distinctive.

Although public transportation is well designed, Los Angeles has one of
the lowest public transportation utilization rates in the United States, and the
mode split of public transportation is only about 5%. The multicenter and
decentralized urban spatial structure makes the passenger volume of the red
line subway too low. The income from tickets accounts for only about 20%
of the total cost of the subway operation.

In fact, the decline of public transportation in Los Angeles began around
the 1950s and 1960s. Before the 1950s, the public transportation in Los
Angeles was relatively developed. In 1946, 426 bus trips were completed per
capita in Los Angeles. However, on the one hand, with the spread and development of the city and the construction of the highway network, public transportation was squeezed, and there were insufficient passengers. On the other
hand, the government did not pay attention to the construction of public
transport. Without funds for public transportation, the lines were forced to
stop or were reduced, especially in the 1940s when the well-built tram was
completely demolished.

At present, there are 183 lines of the Los Angeles ground bus, with a service area of 3711 km^2, with 36 km of bus lanes, and an average of 1.1 million passengers per day. Relevant departments of the Los Angeles
government have also made a great deal of effort to improve the public
transportation utilization rate, and provide shuttle buses for citizens taking
trains. Fig. 20.5 shows buses in front of the Los Angeles railway station.

FIGURE 20.5 A bus in front of the train station in downtown Los Angeles.

FIGURE 20.6 Buses provide convenience for passengers by carrying bicycles.

As shown in Fig. 20.6, buses conveniently allow passengers to carry bicycles, which reflects the efforts being made by Los Angeles to promote the transfer between bicycles and public transport. The rail transit in Los Angeles is also underdeveloped. It has three light rail systems and two underground railways, with a total length of 117.64 km and 62 stations. The daily traffic volume of the subway is about 164,000 day, and the daily traffic volume of the light rail system is about 202,000 per day (2012). The construction of the rail transit system has played a limited role in reducing urban traffic congestion, but due to the car culture and atmosphere together with the spread of low-density development, the private car remains the leading mode of travel in Los Angeles.

20.5   Experiences

Generally speaking, there are more lessons to be learnt than positive experiences in the urban development of Los Angeles, which is worthy of analysis and reference. The low-density spread of urban land and the dominance of cars in the transportation system have brought traffic congestion, high energy consumption, serious environmental pollution, and serious waste of land resources to the city. This chapter summarizes the main experiences and lessons of Los Angeles in urban and urban transportation development, as follows:

1. *Sprawling urban development*

    The low-density traffic demand brought about by the urban sprawl cannot support the operation and development of mass public transport, which leads to a vicious cycle of a low level of public transportation service, citizens turning to cars, further deterioration of the public transportation operation, and a lower level of public transportation service, which leads to the situation that Los Angeles citizens rely too heavily on private cars today.

2. *Unbalanced development of private cars and public transportation*

    The Los Angeles government has failed to find a reasonable balance between the development of cars and public transportation. The development of cars and the construction of highways have been in the ascendance in the second half of the 20th century. The public, entrepreneurs, and their political representatives who support the development of cars have excessively squeezed the development space of public transport, resulting in a situation that the development of public transportation lags far behind the development of cars and highways in general.

3. *A good public transportation culture has not been formed*

    The car culture in Los Angeles has always occupied an important position. The love for cars is a part of the city characteristics of Los Angeles, which has resulted in Los Angeles's originally underdeveloped public transportation system having even less room for improvement. The experiences of Los Angeles show that good guidance of green transportation culture is an important part of a transportation development strategy, and one-sided respect for the private car culture is not conducive to the construction of an urban green transportation system, and will ultimately bring about a very adverse impact on energy use, the environment, and society.

In view of the reflection and re-understanding of the above problems, in recent years, Los Angeles has formulated system planning schemes and many countermeasures, such as strengthening the development of public transport and providing better travel conditions for green travel modes, in order to change the current negative situation of urban traffic in Los Angeles.

Further reading

<http://www.census.gov/>.

Xueming C. Historical evolution of the Los Angeles urban spatial structure. Foreign Urban Plan 2004;19(1):35−41 [in Chinese].

McGuckin N. Travel in LA county; 2009. <http://www.travelbehavior.us>.

Newton D. Density, car ownership, and what it means for the future of Los Angeles; 2010. <https://la.streetsblog.org/2010/12/13/density-car-ownership-and-what-it-means-for-the-future-of-los-angeles/>.

Los Angeles County Metropolitan Transportation Authority. Current situation, problems and challenges of traffic development in Los Angeles. World Metropolitan Transport Development Forum; 2012. [in Chinese].

Wei H, Jiangping Z, Yin X. Government, market and people's preference: experiences and implications of transit development in Los Angeles. Int Urban Plan 2012;27(6):103−8 [in Chinese].

Wikipedia: Los Angeles. <http://zh.wikipedia.org/>.

Index

Note: Page numbers followed by "*f*" and "*t*" refer to figures and tables, respectively.

Printed in the United States
By Bookmasters